SpringerBriefs in Molecular Science

History of Chemistry

Series Editor

Seth C. Rasmussen

Seth C. Rasmussen

How Glass Changed the World

the World

The History and Chemistry of Glass from Antiquity to the 13th Century

 Springer

Seth C. Rasmussen
Department of Chemistry and Biochemistry
North Dakota State University
Fargo, ND
USA

ISSN 2191-5407   e-ISSN 2191-5415
ISBN 978-3-642-28182-2   e-ISBN 978-3-642-28183-9
DOI 10.1007/978-3-642-28183-9
Springer Heidelberg New York Dordrecht London

Library of Congress Control Number: 2012931419

Springer is part of Springer Science+Business Media (www.springer.com)

Acknowledgments

I would first and foremost like to thank the National Science Foundation (CHE-0132886) for initial support of this research and the Department of Chemistry and Biochemistry at North Dakota State University for supporting my continuing efforts in the history of chemistry.

The historical work included in the current volume began as an interest in the early introduction and development of chemical glassware, which led to an initial presentation given at the 227th National Meeting of the American Chemical Society (ACS) in 2004 as part of programming for the Division of the History of Chemistry (HIST). An invitation from E. Thomas Strom then led to an updated presentation as part of the *Perspectives in the History of Chemistry* symposium at the 60th Southwest Regional ACS Meeting later that same year. Continued research finally led to enough information to frame an initial manuscript, which was published in the *Bulletin for the History of Chemistry* in 2008. Throughout this time period, I was warmly welcomed by the members of HIST and strongly encouraged to follow my historical pursuits. As such, I need to acknowledge HIST in providing the environment and encouragement which allowed the development of my initial historical interests into active research and contributions in the history of science. In particular, I wish to acknowledge HIST members David E. Lewis, Carmen J. Giunta, and E. Thomas Strom, as well as Paul R. Jones who was the Bulletin Editor during the submission and publication of my first historical paper.

Of course, the history behind the development of glass and its applications to chemical apparatus continued to hold my interest and research continued, ultimately resulting in this current volume. As part of the preparation of this volume, I would also like to thank my brother, Kent A. Rasmussen, for his most helpful discussions on linguistics, philology, and the intertwining of language and culture. In addition, I would like to acknowledge the Interlibrary Loan Department of North Dakota State University, who went out of their way to track down many elusive and somewhat obscure sources. Lastly, I would like to thank the following current and former members of my research group for reading various drafts of this manuscript and providing critical feedback: Dr. Christopher L. Heth, Michael E. Mulholland, Kristine L. Konkol, Brendan J. Gifford, and Casey B. McCausaland.

Finally, and perhaps most importantly, I must give heartfelt thanks to Elizabeth Hawkins at Springer, without whom this new series of historical volumes would not have become a reality.

Abstract

Glass production is thought to date to ~ 2500 BCE and glass had found numerous uses by the height of the Roman Empire. The modern application of glass to chemical apparatus (beakers, flasks, stills, etc.) was quite limited, however, due to a lack of glass durability under both rapid temperature changes and chemical attack. In the mid-1200s, this began to change as the glassmakers of Venice and Murano began blending previous Roman methods with raw materials from the Levant, as well as developing pretreatment and purification methods of the raw materials used. The combination of these practices resulted in a new glass with a strength and high melting point suitable for use in chemical apparatus. The ability to produce vessels from glass allowed much greater freedom and versatility in the design of laboratory glassware. The resulting improved glass technology led to the invention of eyeglasses, significantly extending the intellectual lifespan of the average scholar. In addition, the freedom of design provided by glass resulted in a vast improvement in still design, which in turn allowed the isolation of important species such as alcohol and the mineral acids. This text provides an overview of the history and chemistry of glass technology from its origins in antiquity to its dramatic expansion in the thirteenth century, then concluding with its impact on society in general, particularly its effect on chemical practices.

Keywords Glass · Pyrotechnology · Materials science · Laboratory equipment/apparatus · Eyeglasses/lenses

Contents

Chapter 1
Introduction

Glass[1] and its uses predate recorded history. Long before the ability to manufacture glass, early tribes discovered and shaped glass formed by nature. Such dark volcanic glass, or *obsidian*, is a naturally occurring silica-based material which is formed from the rapid cooling of volcanic lava (Fig. 1.1). Obsidian can be found in most locations that have experienced the melting of silica-rich rock due to volcanic eruptions and such deposits were valued by prehistoric tribes due to the fact that it could be fractured to produce sharp blades or arrowheads [1].

The technology of synthetic glass production, however, is thought to date back to no later than 3000 BCE [2–5]. This glass technology was not discovered fully formed, but grew slowly through continued development of both chemical composition and techniques for its production, manipulation, and material applications. This development had become fairly advanced by the Roman period, and the first to fourth century CE is often described as the *First Golden Age of Glass* [1, 6]. The manufacture and use of glass became more widespread during the Roman Empire than it had been at any other previous time in history, and glass manufacture flourished in every country under Roman rule (Egypt, Syria, Greece, Italy, and the western provinces of Gaul and Brittany) [1, 7]. During this time, glass was widely used for blown vessels, pitchers, bottles, jars, cups, goblets, bowls, plates, and

---

[1] The modern word '*glass*' used throughout this volume derives from the Old English '*glæs*' with its respective origin in the old West Germanic word '*glasam*', which in turn, is believed to derive from the proto-Indo-European (ca 3500 BCE) root '**ghel*', meaning "to shine or glitter". The earliest known words for glass, however, originate in Mesopotamia with the Akkadian word '*mekku*' and Hurrian word '*ehlipakku*' [10–12]. Both words were also used later in Egypt when glass was introduced there [10, 11]. In Latin, the word for glass is '*vitrum*', from which the modern word '*vitreous*' originates. While absent from known literature before 70 BCE, its use was common after this point [13]. Thus in the Roman period, references to glass used '*vitrium*' and its combining form '*vitri*', as exemplified by the first century historian Pliny the Elder in his *Naturalis Historia* [14].

S. C. Rasmussen, *How Glass Changed the World*,
SpringerBriefs in History of Chemistry, DOI: 10.1007/978-3-642-28183-9_1,
© The Author(s) 2012

Fig. 1.1 Snowflake obsidian

other tableware, with such glass objects becoming as widespread as pottery [1–4, 7, 8]. By the third century, evidence of window glass began to emerge in the writings of Roman authors and in the fourth century, glass use had developed to the point that certain kinds of glass were actually considered a household necessity, although many still remained luxury items [1].

Glass was unlike any other material of this time period and its closest modern material analogues are the organic plastics utilized extensively today. Molten glass can be poured into almost any shape and retains that shape upon cooling, making it an extremely versatile material. While liquid at high temperature, glass is characterized as a *supercooled liquid* or *amorphous solid* at room temperature [7, 9]. That is, glass is a solid, but due to its disordered nature, has properties similar to a liquid that is too viscous to flow at room temperature. Most glass is comprised predominately of silica [6], with an empirical formula of SiO_2. The most common form of natural silica is quartz. However, quartz is a crystalline solid and has a regular repeating crystalline lattice, while glass has no regular repetition in its macromolecular structure and therefore has a disordered structure much like a substance in the liquid state [7]. Silicate solids, both crystalline and glass, are extended three-dimensional networks built up of SiO_4 tetrahedra (Fig. 1.2a) in which the oxygens of adjoining tetrahedra share corners with each other in such a manner that an oxygen atom is linked between two silicon atoms (Fig. 1.2b). For clarity, simplified two-dimensional examples of crystalline and disordered forms of SiO_4 networks are illustrated in Fig. 1.2.

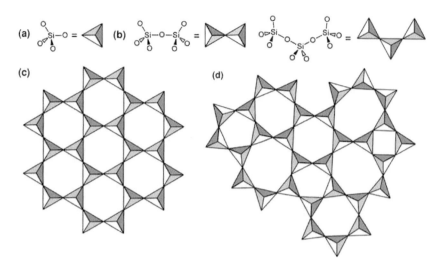

Fig. 1.2 Simplified silicate structures: basic SiO_4 tetrahedra (**a**); short oxygen-bridged units (**b**); two-dimensional crystalline structure (**c**); two-dimensional amorphous structure (**d**)

Due to the extent of disorder in the amorphous structure, the energy of the glass state is typically higher than that of the corresponding crystalline state. The conclusion that glasses represent a metastable state of increased energy is supported by the fact that many glasses tend to undergo devitrification (i.e. glass crystallization resulting in frosting and loss of transparency). Devitrification then represents a reordering of the structure to obtain the more thermodynamically stable crystalline state. Glasses which do not devitrify very rapidly are generally only slightly higher in energy than the corresponding crystalline structure. Thus, the formation of a stable glass occurs when a substance can form a disordered, three-dimensional extended network, which is of comparable energy to the corresponding crystalline state [9].

The source of silica used as the primary component of glass typically came from either beach sand or quartz pebbles [4, 7, 15]. While the fusion of such silica sources alone can result in a glass state, the temperature needed to melt silica (~ 1710°C) was too high to be achieved via the methods available to ancient craftsmen [16, 17]. This problem was solved sometime around 2000 BCE, when craftsmen in Egypt or Mesopotamia discovered that the melting point of silica could be significantly lowered by the use of a *flux* (from the Latin *fluxus*—"flow") [16]. The origin of the flux derives from metallurgy, where it consisted of a component added to the smelting furnace to render the slag liquid at the smelting temperature [18]. As the slag of the smelting furnace consisted primarily of a mixture of silicates and aluminates (rock, sand, and other minerals), its application to the fusion of silica was a logical progression and points to a strong connection between metallurgy and genesis of glassmaking. The most common flux employed by early glassmakers was soda (sodium carbonate or Na_2CO_3) [16], and its application could reduce the melting point of the silica to below 1000°C [17].

While the application of soda significantly lowered the temperature needed to produce molten glass, the high solubility of the sodium contained in the resulting glass tended to make it susceptible to attack by water, thus producing a glass of low chemical stability. To counter this, a third component was needed which would act as a stabilizer for the glass product. Stabilizing species consisted of less soluble ions, particularly calcium or magnesium, from sources such as lime or other mineral additives [16]. The importance of lime, however, was not initially recognized and it was not intentionally added as a major constituent before the end of the seventeenth century [19, 20]. Prior to that time, all calcium content in the ancient glasses was a result of impurities in either the silica or soda sources.

The addition of combinations of soda and calcium salts to the silica can result in an even greater reduction in the mixture's melting point, with temperatures as low as ∼725°C. The triple eutectic mixture providing the lowest melting temperature results in a finished glass of the composition 21.3% Na_2O, 5.2% CaO, and 73.5% SiO_2 [16]. The product of this combination has thus been called soda-lime glass, and most early glasses in the western world were soda-lime-silica compositions that varied depending upon the availability of raw materials [7].

Commonly, these materials consisted of beach sand and a crude source of soda, with both the sand and the soda containing enough lime or magnesia to give some chemical stability [4]. When these species were well mixed and then heated, the soda would almost immediately begin to fuse. This fused material would then begin to react with the sand to generate sodium silicates and initial liquid material. Lime and other bases would then begin to combine with the silica and join the melt, with the excess silica being the last to fuse as the melt temperature increases and the viscosity is reduced [21]. Throughout this process, gases would be liberated as the various carbonate, nitrate, and sulfate components are converted to the respective oxides, resulting in violent agitation of the mixture and the generation of significant bubbles in the final melt [21]. In order to reduce the presence of bubbles in the final glass, this process was typically accomplished in two distinct stages. In the first stage, the mixture was heated at a temperature to allow the reaction of the soda and lime with the silica, and thus liberate the majority of the gaseous byproducts, but not hot enough to achieve homogenous fusion of the complete mixture. This intermediate product is referred to as a *frit* or preliminary sintered mixture [22]. The intermediate frit was then crushed to enhance intimate mixing and heated a second time at increased temperature to generate the final glass relatively free of bubbles [22].

Of course, early glass contained a wide variety of unintended impurities besides calcium or magnesium, and these additional components influenced the properties of the resulting glass in various ways. One such effect of impurities was the resulting color of the glass. Early glass produced in antiquity was rarely colorless [1], primarily due to impurities of iron in the sand or flux. All sands contain a certain amount of Fe_2O_3, which is a very strong coloring agent and commonly gives a light blue-green color to the glass [7, 16, 17]. While the coloring effect of

Table 1.1 Coloring agents for the generation of colored glass in soda-lime glasses

Color	Transition or main group metal	Resulting oxides responsible for color	References
White	Calcium plus antimony, or tin	$Ca_2Sb_2O_7$; SnO_2	[1, 7, 16, 23, 26]
Yellow	Lead plus antimony, or iron	$Pb_2Sb_2O_3$; Fe_2O_3	[1, 7, 16, 23, 25]
Orange	Chromium	CrO_3	[25]
Red	Copper or lead	Cu^a; Pb_3O_4	[1, 7, 16, 26]
Purple	Manganese	Mn_2O_3	[1, 7, 23, 25, 26]
Brown-violet	Nickel	NiO	[25]
Blue	Cobalt	CoO	[1, 7, 16, 23, 25, 26]
Blue-green	Iron or copper	FeO, CuO	[1, 7, 16, 25, 26]
Green	Chromium	Cr_2O_3	[25]

[a] The formation of metallic copper nanoparticles can result in a ruby red color

the iron might have been initially unintentional, early glassmakers developed a number of transition metal additives to generate a variety of colored glasses (Table 1.1). These transition metal salts would generate metal oxides during the formation of the molten mixture, providing strong coloring of the resulting glass. Green or red could be produced from the addition of various copper species, blue from cobalt, blue-green from copper or iron, and yellow from a combination of lead and antimony. Such colored glass was normally transparent, but the addition of sufficient lead, antimony, or tin could make the glass opaque [7, 15, 23]. The application of such colored glass to stained-glass windows dates back to at least the tenth century [24].

The final color of the glass would depend not only on the metal oxide used, but also on the amount of colorant used and the nature of the glassmaking, particularly the chemical composition of the flux and the manner of heating [25, 26]. For example, the application of nickel oxide in soda-lime glass would result in a brownish violet color. However, changes in either the flux or the stabilizer could result in colors ranging from yellow to purple [25]. Black glass can be made by using almost any of the more powerful colorants in sufficient concentration. In these cases, the glass is so dark that it appears black, although very thin sections will reveal it to be the color of the chosen colorant. For example, the application of high quantities of manganese can result in a black glass, which is seen to be actually a very dark purple in sufficiently thin glass samples [25, 26].

Colorless glass was finally achieved by the careful selection of fine sand of low iron content and by the addition of decolorizing agents such as manganese dioxide or antimony [16, 27]. Most decolorizing agents served to oxidize the strongly absorbing Fe(II) species to the fairly colorless Fe(III) to inhibit the coloring properties of the iron impurities [16]. Colorless glass produced in this way is transparent unless it contains air bubbles or undissolved particles. The majority of ancient glass, however, contained much undissolved material and thus was not as transparent as modern glass.

While early glass is typically referred to as soda-lime glass, the discussion above has shown that it can be much more complex than this simple name

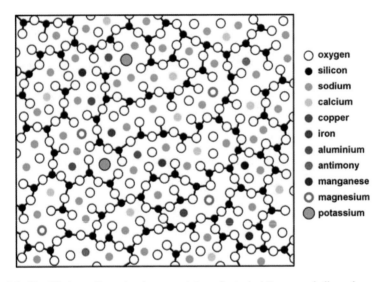

Fig. 1.3 Simplified two-dimensional representation of a typical Roman soda-lime glass

suggests. In addition to the three primary components of silica, flux, and stabilizer, the majority of such glasses also contained one or more coloring or decolorizing agents, as well as a fair number of impurities introduced along with each of the intended glass constituents. As a consequence, the chemical composition and structure could be both quite complex and extremely variable, resulting in glasses with a range of chemical and physical properties. In addition, the character of glass depends not only on the raw materials used, but also the manner and degree of heating, as well as the method and rate of cooling the hot glass (annealing) [28]. The complexity of such soda-lime glasses is highlighted in Fig. 1.3, which provides a simplified, two-dimensional illustration of the composition and structure of a typical Roman glass.

In addition to the various applications of glass already discussed, it has come to be used extensively in modern science, particularly in terms of laboratory glassware. In fact, the image of glass objects such as distillation heads, beakers, flasks, vials, and test-tubes has become mainstay in the public's common view of chemical laboratories and the practice of chemical research. This common view is illustrated by the use of such images to depict various aspects of chemical research on various postage stamps as shown in Fig. 1.4. Even for the practitioner of science, it is hard to imagine what the chemical laboratory would be like without glass. After all, no other material can really match its combination of low cost, chemical stability, thermal durability, and transparency, as well as freedom and versatility in the design of chemical apparatus for nearly any needed application. Although glass and its applications were widespread throughout the Roman Empire, the use of glass for laboratory apparatus prior to the thirteenth century was still severely limited because of a lack of durability under rapid temperature

Fig. 1.4 Various postage stamps depicting laboratory glassware [Courtesy of Dan Rabinovich]

changes and poor chemical resistance. For example, the combination of poor quality and the thick, irregular nature of the glass resulted in the frequent breaking of the vessels during distillation [29]. This changed in the thirteenth century when further advances in the production of glass were developed in Venice and Murano, ushering in the *Second Golden Age of Glass* [1].

The present volume is an extension of previous work [30] with the goal to bring together various partial works in history, chemistry, archaeology, and glass studies. Such a combined approach can provide a more detailed picture of the development of glass from its beginnings in antiquity to its dramatic expansion in the thirteenth century, resulting in its full use as a commonplace material throughout society. While the early history of glass has been described previously by a number of authors, the focus has been primarily on glassmaking as a technology or art form and less emphasis has been given to the rich and diverse chemistry involved. Of particular interest, this current report aims to provide an understanding of how and why the Venetian glass of the thirteenth century became suitable for use in chemical apparatus and what effect the availability of this new glass had on the long-term progress of laboratory practitioners. For example, it could be argued that the improved glass technology of Venice led to the invention of eyeglasses and a vast improvement in still design. Both events occurred shortly after the introduction of the improved glass and within close geographical proximity to Venice. As a result of newly available glass-based chemical labware, particularly more advanced distillation apparatus, important materials were isolated in pure forms for the first time, most notably alcohol and the mineral acids. The availability of these materials then greatly affected the evolving fields of both chemistry and medicine, which in turn changed the world as we know it. This lasting power and impact of glass was eloquently stated by eighteenth century author Samuel Johnson, saying [31]:

Who, when he saw the first sand or ashes, by a casual intenseness of heat, melted into a metalline form, rugged with excrescences, and clouded with impurities, would have imagined, that in this shapeless lump lay concealed so many conveniences of life, as would in time constitute a great part of the happiness of the world? Yet by some such fortuitous liquefaction was mankind taught to procure a body at once in a high degree solid and transparent, which might admit the light of the sun, and exclude the violence of the wind; which might extend the sight of the philosopher to new ranges of existence, and charm him at one time with the unbounded extent of the material creation, and at another with the endless subordination of animal life; and, what is yet of more importance, might supply the decays of nature, and succour old age with subsidiary sight. Thus was the first artificer in glass employed, though without his own knowledge or expectation. He was facilitating and prolonging the enjoyment of light, enlarging the avenues of science, and conferring the highest and most lasting pleasures...

References

1. Philips CJ (1941) Glass: the miracle maker. Pitman Publishing Corporation, New York, pp 3–14
2. Cummings K (2002) A history of glassforming. A & C Black, London, pp 102–133
3. Macfarlane A, Martin G (2002) Glass, a world history. University of Chicago Press, Chicago, pp 10–26
4. Kurkjian CR, Prindle WR (1998) Perspectives on the history of glass composition. J Am Ceram Soc 81:795–813
5. Morey GW (1954) The properties of glass, 2nd edn. ACS monograph series No. 124, Reinhold Publishing, New York, pp 5–6
6. Axinte E (2011) Glass as engineering materials: a review. Mater Des 32:1717–1732
7. Brill RH (1963) Ancient glass. Sci Am 109(5):120–130
8. McCray WP, Kingery WD (1988) Introduction: toward a broader view of glass history and technology. In: McCray WP, Kingery WD (eds) The prehistory and history of glassmaking technology ceramics and civilization VIII. American Ceramic Society, Columbus, pp 1–13
9. Zachariasen WH (1932) The atomic arrangement in glass. J Am Chem Soc 54:3841–3851
10. Shaw I, Nicholson PT (1995) British museum dictionary of ancient Egypt. British Museum Press, London, pp 112–113
11. Cummings K (2002) A history of glassforming. A & C Black, London, p 48
12. Saldern AV, Oppenheim AL, Brill RH, Barag D (1988) Glass and glassmaking in ancient mesopotamia. Corning Museum of Glass Press, Corning, pp 56–57, 78
13. Lambert JB (1997) Traces of the past: unraveling the secrets of archaeology through Chemistry. Addison-Wesley, Reading, p 113
14. Pliny the Elder (1906) Naturalis Historia. Mayhoff KFT (ed) Teubner, Lipsiae, Book XXXVI
15. Jacoby D (1993) Raw materials for the glass industries of venice and the terraferma, about 1370—about 1460. J Glass Studies 35:65–90
16. Lambert JB (2005) The 2004 Edelstein award address, the deep history of chemistry. Bull Hist Chem 30:1–9
17. Philips CJ (1941) Glass: the miracle maker. Pitman Publishing Corporation, New York, pp 36–44
18. Raymond R (1986) Out of the fiery furnace. the impact of metals on the history of mankind. The Pennsylvania State University Press: University Park, pp 16, 27
19. Turner WES (1956) Studies in ancient glasses and glassmaking processes. Part III. The chronology of the glassmaking constituents. J Soc Glass Technol 40:39T–52T

20. Turner WES (1956) Studies in ancient glasses and glassmaking processes. Part V. Raw materials and melting processes. J Soc Glass Technol 40:277T–300T
21. Philips CJ (1941) Glass: the miracle maker. Pitman Publishing Corporation, New York, pp 152–153
22. Saldern AV, Oppenheim AL, Brill RH, Barag D (1988) Glass and glassmaking in ancient mesopotamia. Corning Museum of Glass Press, Corning, pp 112–113
23. Lambert JB (1997) Traces of the past. unraveling the secrets of archaeology through chemistry. Addison-Wesley, Reading, p 105
24. Roger F, Beard A (1948) 5,000 years of glass. J. B. Lippincott Co., New York, pp 157–176
25. Philips CJ (1941) Glass: the miracle maker. Pitman Publishing Corporation, New York, pp 48–51
26. Roger F, Beard A (1948) 5,000 years of glass. J. B. Lippincott Co., New York, pp 108–111
27. Tait H (ed) (1991) Glass, 5,000 years. Harry N. Abrams, Inc., New York, p 22
28. Roger F, Beard A (1948) 5,000 years of glass. J. B. Lippincott Co., New York, pp 1–5
29. Forbes RJ (1970) A short history of the art of distillation. E. J. Brill, Leiden, pp 76–85, 114
30. Rasmussen SC (2008) Advances in 13th century glass manufacturing and their effect on chemical progress. Bull Hist Chem 33:28–34
31. Johnson S (1826) The rambler. Thomas Tegg, London

Chapter 2
Origins of Glass: Myth and Known History

Where and when glass production began is uncertain. It is thought by some that the first glass was probably developed in the Mitannian or Hurrian region of Meso-potamia, possibly as an extension of the production of glazes ($\sim$ 5000 BCE) [1]. Around this same time, a new material called *faience* was developed, which was produced by utilizing a variety of techniques to create a glaze layer over a silica core [2, 3]. It may have been invented in either Sumeria or Egypt, but its full development was accomplished in Egypt, and it is therefore commonly referred to as Egyptian faience [2]. Although this material was used to craft beads during the third and fourth millennia BCE, it involved sintering (fusion below the melting point), rather than the complete melting of the silica mixture [4]. As such, faience can be thought of as an intermediate material between a glaze and glass [4]. Glass as an independent material is not thought to predate 3000 BCE, with the first glass objects including beads, plaques, inlays and eventually small vessels [1, 5–7]. Glass objects dated back to 2500 BCE have been found in Syria, and by 2450 BCE, glass beads were plentiful in Mesopotamia [4]. Glass came later in Egypt, with its manufacture appearing as a major industry around 1500 BCE [4, 8–11]. The oldest glass of undisputed date found in Egypt dates from $\sim$ 2200 BCE [12].

S. C. Rasmussen, *How Glass Changed the World*,
SpringerBriefs in History of Chemistry, DOI: 10.1007/978-3-642-28183-9_2,
© The Author(s) 2012

2.1 Myth and Legend

Many legends have attempted to explain the discovery of glassmaking. The most famous of these was recorded by the first century historian Pliny the Elder[1] in his *Naturalis Historia* (*Natural History*) [13]:

> In Syria there is a region known as Phœnice, adjoining to Judæa, and enclosing, between the lower ridges of Mount Carmelus, a marshy district known by the name of Cendebia. In this district, it is supposed, rises the river Belus, which, after a course of five miles, empties itself into the sea near the colony of Ptolemaïs. The tide of this river is sluggish, and the water unwholesome to drink, but held sacred for the observance of certain religious ceremonials. Full of slimy deposits, and very deep, it is only at the reflux of the tide that the river discloses its sands; which, agitated by the waves, separate themselves from their impurities, and so become cleansed. It is generally thought that it is the acridity of the sea-water that has this purgative effect upon the sand, and that without this action no use could be made of it. The shore upon which this sand is gathered is not more than half a mile in extent; and yet, for many ages, this was the only spot that afforded the material for making glass.
>
> The story is, that a ship, laden with nitre,[2] being moored upon this spot, the merchants, while preparing their repast upon the sea-shore, finding no stones at hand for supporting their cauldrons, employed for the purpose some lumps of nitre which they had taken from the vessel. Upon its being subjected to the action of the fire, in combination with the sand of the sea-shore, they beheld transparent streams flowing forth of a liquid hitherto unknown: this, it is said, was the origin of glass.

Pliny's account places the discovery of glass in the north of modern Israel, just south of Lebanon (Fig. 2.1). The Belus river is identified with what is now known

---

[1] Pliny the Elder or Gaius Plinius Secundus (23–79 CE) was a Roman officer and encyclopedist. He was born in late 23 or early 24 at Novum Comum (modern Como), a small city in the region known as Transpadane Gaul (or Gallia Transpadana). Introduced to the city of Rome at an early age, he studied there before going on to become a military tribune at age 21. As an army officer, he held three posts, at least two of which were served in Germany. Best known as a writer and encyclopedist, he wrote his first treatise in 50–51, followed by a two-volume biography of the senator *Pomponius Secundus* and the twenty-volume *History of Rome's German Wars*. Following this, his writing shows a change in direction, thought to be associated with his final return to civilian life. He is most well-known for his encyclopedia, *Naturalis Historia*, published in 77 CE. This massive work resulted from years of collecting records, both from his reading and from personal observations, or anything and everything that seemed to him worth knowing. He died in late August of 79 during the evacuation around the erupting volcano Vesuvius. The exact cause of his death is unknown, but it has been said that he was asthmatic and overcome by sulfurous fumes. It is reported that he was still recording the personally observed marvels of nature to the last hours of his life [14].

[2] Alkaline carbonate, typically soda (sodium carbonate or Na_2CO_3). The word '*nitre*', which most recently refers to sodium nitrate, has only acquired that meaning within recent centuries. Originally it meant carbonated alkali, something that effervesced with vinegar or other acid, and when dissolved in water was a cleansing agent. The ancient Egyptians obtained native soda called '*nitrike*' from lakes such as those in Nitria. The Greek word became '*nitron*' and in turn became the Latin '*nitrum*' and the European '*nitre*'. Thus, the Greek '*nitron*' used by Hippocrates in the fifth century BCE, the Latin '*nitrum*' of Pliny in the first century CE, and their English equivalent '*nitre*', all apply to the soda obtained from either evaporitic lakes or plant ash [16].

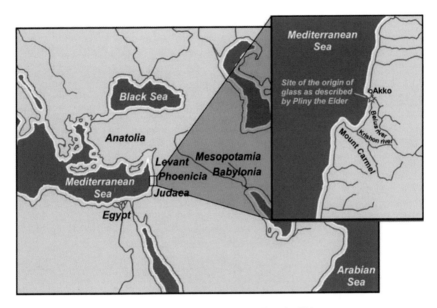

Fig. 2.1 Site of the discovery of glass as described by Pliny the Elder

as the Na'aman river, and the mouth of the Belus resided just south of the city of Akko (modern Acre). Analysis of the sand at the mouth of the Belus has revealed that it indeed is a high silica sand containing sufficient quantities of calcium components, yet with little other measureable impurities [15]. In addition to Pliny, the Belus sand has been referred to by a number of classical writers and is thought to have served as a long-time silica source for glassmakers working along the Syrian coast. Its exportation to other glassmaking centers has also been proposed [15].

In *The Art of Glass*, seventeeth century glassmaker Antonio Neri[3] gives a slightly different account, although again occurring at the mouth of the Belus river. Neri credits this tale again to Pliny the Elder [17]:

> Pliny saith, that Glass was found by chance in Syria, at the mouth of the river Bellus, by certain Merchants driven thither by the fortune of the Sea, and constrained to abide there and to dress their provisions, by making fire upon the ground, where was great store of this sort of herb which many call Kali, the ashes whereof make Barilla, and Rochetta; This herb burned with fire, and therewith the ashes & Salt being united with sand or stones frit to be vitrified is made Glass.

---

[3] Antonio Neri (d. 1614) is the author of the Italian manuscript *L'Arte Vetraria* (*The Art of Glass*). Initially published in 1612, it is considered to be the world's most famous book on glassmaking. Little is actually known about Neri, but he has been generally referred to as a Florentine monk and the tone of this writing is consistent with this profession. He was but one of several monks who, over a period of several centuries, were important contributors to the knowledge of glass [25].

To investigate the possibility of the discovery of glass as described in these accounts, William L. Monro of the American Window Glass Co. attempted to recreate the conditions described during a series of experiments in the 1920s [18]. Over a bed of glass sand mixed with an equal quantity of carbonate of soda, he built an open wood fire which he kept burning for two hours. As the fire burned, he monitored the temperature generated using a standard pyrometer couple inserted into the bed of the fire. He determined that a maximum temperature of 2210°F ($\sim$1210°C) was obtained when the fire had been reduced to a mass of burning charcoal. After the fire had completely burned itself out, the ashes were removed and a portion of the bed was found to be fused into a vitreous mass.

He then repeated the process, this time using a bed of glass sand mixed with an equal quantity of nitre.[4] As before, a portion of the bed mixture was found to have undergone fusion [18]. Finally, he carried out the process a third time, now using only the bed of glass sand, unmixed with any other ingredients. In this last case, examination of the sand bed after removal of the ashes revealed no evidence of even the slightest trace of fusion.

While Monro felt that these results confirmed the plausibility of the Pliny's story [18], others have pointed out some important considerations. The first consideration is that in Munro's recreation, a large quantity of soda was mixed throughout the sand, rather than the relatively limited interface of soda and sand described by Pliny [19]. As such, Munro's conditions were much more favorable for the production of the fused products and are not truly an accurate recreation of Pliny's story. The second point made is that it can be assumed that the merchants were interested in a fire hot enough to cook, not necessarily the extreme temperature achieved by Monro [18]. In fact, it has been reported by multiple sources that an ordinary campfire does not reach much higher than 600–650°C [20], with a possible maximum of 700°C [21–23]. As such, the temperatures claimed by Monro are significantly high and it is unclear exactly how these extreme temperatures were achieved. In fact, as the fusion temperature of a one-to-one mixture of sand and soda is typically below 1000°C [24], the claimed temperature of 1210°C should have resulted in fusion of a greater portion (if not all) of the bed. Nevertheless, even with the conditions tipped in his favor, Monro did not observe the free-flowing liquid glass described by Pliny.

Munro goes on to mention that seaweed ash contains a large amount of sodium carbonate and has been used in glassmaking to produce what has been called 'kelp glass'. He thus states [18]:

> It requires no great stretch of the imagination to think that at some time there had been kindled along a sandy shore a great bonfire of dry seaweed, with perhaps a lot of driftwood, which left amid its charred embers the vitreous mass we now call glass.

While such conditions proposed by Munro are speculation, they do follow fairly closely with Neri's account above.

---

[4] While it is unclear, it appears that sodium nitrate ($NaNO_3$) was used in the second attempt. It may be that Munro is interpreting Pliny's use of 'nitre' in its modern sense here.

2.2 Current Historical Knowledge

While the accounts discussed so far make entertaining stories, they are not commonly accepted as historically accurate and currently scholars believe that glass was discovered either as a byproduct of metallurgy or from an evolutionary sequence in the development of ceramic materials [11, 21]. These two hypothetical origins are deemed plausible as both early technologies had procedures that could be considered precursors of glass [4]. Considering the possibility that glass arose from metallurgical operations, a brief discussion of the history of metallurgy is required. It is known that the smelting of copper began as early as 6000 BCE in Anatolia (modern Turkey) [4]. Others, however, credit the Sumerians in southern Mesopotamia with the origin of copper smelting. By 3700 BCE, copper was being produced in the Sinai Peninsula and a little later ($\sim$ 3000 BCE) on Cyprus, from which the word copper is derived [21].[5]

The smelting of copper consisted of heating the ore malachite ($Cu_2CO_3(OH)_2$) in the presence of charcoal at temperatures of $\sim$ 1200°C [20]. The incomplete combustion of the charcoal would result in a strong reducing atmosphere of carbon monoxide, which would reduce the Cu(II) of the ore to metallic copper. At the temperatures employed, the metallic copper produced would become molten (Cu mp = 1084°C) and could be isolated and cooled to generate pure copper cakes [20, 25]. Of course, a complication in this process is that in collecting the ore, a good deal of rock was unavoidably collected as well. Common rock is comprised of various silicates and aluminosilicates which do not easily melt at the temperatures applied for the smelting of copper. Thus their presence would result in the isolation of a solid heterogeneous mixture of rock and raw metal, which would then have to be broken up and the metal removed, making its isolation cumbersome [25].

To overcome this complication, a flux would be added to assist with the melting of the residual silicate and aluminosilicate species. Early common fluxes for copper smelting were easily fusible pyrites and evidence has been found confirming such iron ores as flux in copper smelting [20, 25]. However, the types of species utilized as fluxes were quite diverse and in addition to various metal ores, also included a number of simple carbonate, sulfate, and nitrate salts. Known examples of such fluxes include soda (Na_2CO_3), potash (K_2CO_3), saltpeter (KNO_3), and vitriols (metal sulfates) [26]. Application of the flux would then result in a combination of molten metal and a fused mixture of rock and flux, commonly referred to as the *slag*. As the molten metal and slag were not miscible, they would form two separate molten layers within the smelting furnace. The two layers could be separated from one another via a process called *liquation*, in which the layers were poured or drained off one layer at a time (Fig. 2.2).

---

[5] The modern term '*copper*' derives from the Old English '*coper*' with its respective origin in the Latin '*cuprum*'. Cuprum in turn is a Roman contraction of '*aes cyprium*', meaning "metal of Cyprus".

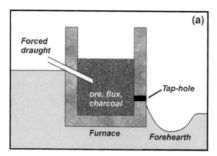
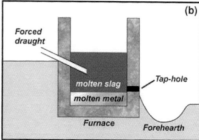
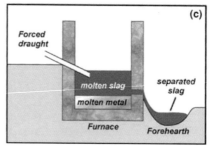

Fig. 2.2 Smelting process: ore, flux, and charcoal are mixed in the smelting furnace and fired (**a**); heating produces immiscible layers of molten metal and slag (**b**); the tap-hole is removed, allowing the slag to drain off into the forehearth (**c**)

When the molten slag was allowed to cool, it produced a rigid, glassy solid similar to obsidian. It is easy to imagine that experimentation with such siliceous slags (variation in types and source of rock, variation in flux, etc.) might well have led to the direct formation of colored vitreous silicate objects [11, 27]. Support for this proposed origin for glassmaking has also included the fact that many early glazes and glasses were colored blue by the addition of copper [21, 27]. However, the most significant evidence for a relationship between ancient glassmaking and metallurgy comes from archaeological finds. The Ramesside Egyptian site of Qantir (late second millennium BCE), contains evidence for both the preparation of red opaque glass ingots and bronze casting in a single site. Thus, this provides a clear example of the production of colored glass taking place at the very site where metallurgical byproducts were being generated [27].

Additional support for this connection comes from the analysis of second millennium BCE light blue opaque Malkata glasses, which revealed the presence of tin oxide [27]. As these glasses are colored with copper species, the presence of tin indicates the potential use of bronze dross, scale, or corrosion products as the source of copper(II) ions to color the glass. Similar relationships have also been observed between the copper and tin contents of blue New Kingdom glasses, the ratio which is compatible with the compositions of New Kingdom bronzes [27].

Of course, it has also been pointed out that slags from copper smelting actually contain only a little copper and are much richer in iron than either the early glazes or glasses [21]. It must be remembered, however, that only very small amounts of copper would be needed to provide the blue color. In addition, the high amount of iron is not surprising considering the common flux for copper was iron pyrites. The move to another flux via experimentation could easily have resulted in early blue glass with low iron content.

The second possible origin for the discovery of glass is thought to be due to an evolutionary development of a family of highly siliceous ceramics coated with alkali glazes originating in either Sumeria or Egypt [2, 11, 21]. The immediate predecessor of glass in this developmental sequence is the material known as faience [11, 21]. Faience was used mostly to make small objects such as beads and is found in profusion at archaeological sites in Egypt and elsewhere [4, 21]. It is produced utilizing a variety of techniques to create a glaze layer over a silica core [2, 3]. The resulting surface of faience is a transparent glass, usually blue or green, encapsulating a body consisting of crystalline grains of quartz loosely bound together by a glassy phase. In some specimens a thin layer of powdered material lies between the glaze and the body [20].

Chemical analysis has shown that the body of faience consists primarily of silica with small amounts of soda and other impurities [11, 21]. The study of several specimens by X-ray diffraction has revealed that the grains of silica consist of α-quartz, indicating that the material was heated to a temperature no higher than 870°C. The application of higher temperatures would have produced domains of tridymite in the body [21]. The formation of faience objects has been easily duplicated in the laboratory. Finely powdered quartz is combined with aqueous sodium carbonate to produce a firm paste, which can then be formed and fired. During the heating, the sodium fuses with the surface of the quartz grains, giving rise to both a glass exterior and an interior glassy phase that binds together the domains of α-quartz. In this case, the crystalline domains predominate, with only a small amount of glassy material and a large proportion of empty space [21].

From this knowledge, it is clear that the initial discovery of glass could have occurred via a few simple variations in the production of faience. Such variations could easily have occurred accidentally due to poor compositional or temperature control (i.e. excess soda or heat), or else as a result of investigating the effect of variable conditions on faience production [21, 28]. For example, if the ratio of sodium carbonate to powdered quartz had been increased in the initial paste, or if the formed paste had been fired at either a higher temperature or for an elongated period of time, the fusion of quartz and soda could have proceeded to a greater extent. Under such modified conditions, the domains of α-quartz would have been fewer in number and of smaller size, so that the material would have been mostly glassy. Having once made such a crudely formed glass, the faience makers could readily have gone on with a little additional experimentation to produce a true glass without any crystalline domains [21].

Such a path to the discovery of glass is supported by the fact that there is a known type of faience known as glassy faience. The structure of glassy faience is intermediate between the structure of ordinary faience and that of true glass and could thus be a logical intermediate in the path from faience to glass [21].

Unfortunately, the time period for the introduction of such glassy faience is not well documented and thus it is not certain that it was made before the invention of glass itself [21]. An inconsistency that should also be considered with this theoretical path is that while the full development of faience was accomplished in Egypt (and thus commonly referred to as Egyptian faience) [2], glass is thought to have originated in Mesopotamia and Syria, with its spread to Egypt at a later date [1, 6–10]. As the more advanced and significant faience production occurred in Egypt, it would be logical that the transition from faience to glass would also take place among these Egyptian artisans. Of course, this does not eliminate the possibility that the less advanced faience artisans of Mesopotamia accomplished the more significant advance to glass, while the Egyptian craftsman continued to perfect the production of faience without the transition to the new material.

In attempting to explain the delay of more than 2000 years between the production of faience and that of glass, it has been suggested that an important factor was that the production of faience involved only cold-working and reduced temperature sintering of the raw materials [4, 28]. In contrast, the routine production of glass vessels and other objects involved the manipulation of hot, viscous fluids, a process that was more akin to metal working. Therefore, although the production of glazed stones, faience, and glass all involved the same combination of essentially identical raw materials, the change from cold-working for glazes and faience to hot-working for glass may not have been a logical progression or an easy transition [28].

Such a transition would most likely have required input from metal workers who were more familiar with such high temperature manipulations. Thus, it can be argued that the discovery of the techniques necessary for hot-working glass was the result of interaction between the workers of glazed stone and faience and metal workers [27, 28]. Further, it is possible that such interactions were a result of the changing control over and organization of artisans following the political upheavals occurring in Egypt and the Near East during the sixteenth century BCE. As a result, artisans skilled in different crafts could have been brought into close proximity in workshops and production centers. In such an environment, the transfer of technologies between crafts would have been facilitated, paving the way for the eventual discovery of glass production [28].

While arguments can be made for either of the two commonly proposed pathways to the origin of glass, it is clear that either path is not completely independent of the other. In the first case, metallurgy is thought to originate in the pottery kilns, potentially as a consequence of using metal ores in glazes. In the second case, the high temperatures required for the production and working of glass is thought to have required input from metal workers. As such, it is quite reasonable to propose a combined path in which transfer of knowledge and observation between the two groups of craftsmen resulted in the discovery of glass with origins in both metallurgy and siliceous glazes.

References

1. Cummings K (2002) A history of glassforming. A & C Black, London, pp 102–133
2. Noble JV (1969) The technique of Egyptian Faience. Amer J Achaeol 73:435–439
3. Cummings K (2002) A history of glassforming. A & C Black, London, p 46
4. Lambert JB (2005) The 2004 Edelstein award address, the deep history of chemistry. Bull Hist Chem 30:1–9
5. Macfarlane A, Martin G (2002) Glass, a world history. University of Chicago Press, Chicago, p 10
6. Kurkjian CR, Prindle WR (1998) Perspectives on the history of glass composition. J Am Ceram Soc 81:795–813
7. Shortland AJ, Tite MS (2000) Archaeometry 42:141–151
8. Smirniou M, Rehren TH (2011) Direct evidence of primary glass production in late bronze age amarna Egypt. Archaeometry 53:58–80
9. Tite MS (2004) Glass and related vitreous materials. Proc Intern School Phys Enrico Fermi 154:369–376
10. Saldern AV, Oppenheim AL, Brill RH, Barag D (1988) Glass and glassmaking in ancient mesopotamia. Corning Museum of Glass Press, Corning, pp 21, 67, 79, 109–111, 184
11. Lambert JB (1997) Traces of the past. Unraveling the secrets of archaeology through chemistry. Addison-Wesley, Reading, pp 105–109
12. Morey GW (1954) The properties of glass, 2nd edn. ACS monograph series no. 124, Reinhold Publishing, New York, pp 5–6
13. Pliny the Elder (1855) The natural history. Bostock J, Riley HT (trans). Taylor and Francis, London, Book XXXVI, Chapter 65
14. Reynolds J (1986) The Elder Pliny and his times. In: French R, Greenway F (eds) Science in the early Roman empire: Pliny the Elder, his sources and influence
15. Turner WES (1956) Studies in ancient glasses and glassmaking processes. Part III. The chronology of the glassmaking constituents. J Soc Glass Tech 40:39T–52T
16. Turner WES (1956) Studies in ancient glasses and glassmaking processes. Part V. Raw materials and melting processes. J Soc Glass Technol 40:277T–300T
17. Neri A, Merrett C (2003) The world's most famous book on glassmaking, the art of glass. The Society of Glass Technology, Sheffield, pp 47–48
18. Monro William L (1926) Window glass in the making. American Window Glass Company, Pittsburgh, pp 20–23
19. Roger F, Beard A (1948) 5,000 years of glass. J. B. Lippincott Co., New York, pp 2–3
20. Aitchison L (1960) A history of metals, vol 1. Interscience Publishers, Inc., New York, pp 36–40
21. Brill RH (1963) Ancient glass. Sci Am 109(5):120–130
22. Raymond R (1986) Out of the fiery furnace. The impact of metals on the history of mankind. The Pennsylvania State University Press, University Park, pp 10–16
23. Coghlan HH (1939) Prehistoric copper and some experiments in smelting. Trans Newcomen Soc 20:49–65
24. Philips CJ (1941) Glass: the miracle maker. Pitman Publishing Corporation, New York, pp 36–44
25. Neri A, Merrett C (2003) The world's most famous book on glassmaking, the art of glass. The Society of Glass Technology, Sheffield, pp 6–9
26. Hoover HC, Hoover LH, Acgricola G (1950) De Re Metallica. Dover Publications, Inc., New York, pp 232–239
27. Mass JL, Wypyski MT, Stone RE (2002) Malkata and Lisht glassmaking technologies: towards a specific link between second millennium BC metallurgists and glassmakers. Archaeometry 44:67–82
28. Tite MS (2004) Glass and related vitreous materials. Proc Intern School Phys Enrico Fermi 154:369–376

Chapter 3
Development and Growth of Glass Through the Roman Period

As outlined in the previous chapter, glass as an independent material is not thought to predate 3000 BCE, with its earliest development located in Mesopotamia and Syria [1–4]. Routine glass production is then thought to have started in Mesopotamia around 1550 BCE [5]. It is generally assumed that glass-working was then introduced into Egypt during the reign of Tuthmosis III (1479–1425 BCE) through a combination of glass objects and ingots being imported as tribute [3, 5–7]. Mesopotamian glassmakers were also thought to have been imported into Egypt [6] so that local production of glass in western Thebes was established by the time of Amenophis III (ca. 1388–ca. 1350 BCE) [3]. However, it is less well established whether Egypt initially relied on imported raw materials (in the form of ingots and cullet) that were then worked in Egypt, or whether glass was actually being produced onsite [3, 6]. Nevertheless, evidence supports onsite glass production in Egypt by 1350 BCE and points to Egypt as a primary glass producer during this early period of glass production [5].

Glass manufacture in the Mediterranean soon become a major industry and was extremely successful for the next 300 years [8]. The current model of this industry proposes the existence of only a few primary glass production sites. Glass in the form of ingots was then traded from these primary sites to secondary glass workshops scattered around the Mediterranean, and even extending into the Aegean world [5]. Glass of this initial period (1500–800 BCE) is characterized as a typical soda-lime glass with a high magnesia (3–7%) and potash (1–4%) content [3, 4, 9], which is representative of glass produced or used throughout the Mediterranean area [9]. Commonly, these materials were produced from a mixture of silica (sand or pebbles) and a crude source of alkali (most commonly soda). Both the silica and alkali could then act as sources of lime or magnesia to give the resulting glass some chemical stability [2].

S. C. Rasmussen, *How Glass Changed the World*,
SpringerBriefs in History of Chemistry, DOI: 10.1007/978-3-642-28183-9_3,
© The Author(s) 2012

Table 3.1 Compositions of various sands from ancient sites[a]

Component	Egyptian sites			Syrian sites	
	Amarna	Karnak	Thebes	Haifa	Mouth of Belus[b]
SiO_2	60.46	83.61	72.69	76.40	80.65
Al_2O_3	2.25	1.32[c]	8.18		4.27
Fe_2O_3	1.73		5.60		0.13
CaO	18.86	12.01	4.86	10.73	8.81
MgO	0.83	1.23	2.44	0.75	
Na_2O	0.30		1.21		
K_2O	0.74		1.10		
Water	0.42	1.57	1.04	0.40	
Organic[d]	13.90		1.60	7.80	6.23

[a] Ref. [11], % by weight
[b] Average of three different samples
[c] Combined total of both Al_2O_3 and Fe_2O_3
[d] By loss on ignition

3.1 Silica Sources

The two primary sources of silica usually cited for this period are quartzite pebbles and sand [10–12]. A study of Assyrian cuniform texts has revealed a raw material called '*immanaku*', which has been interpreted as quartzite pebbles, probably collected from river beds [3, 7]. Analysis of quartzite pebbles has revealed them to be a very pure form of silica, containing only trace impurities of barium (19 ppm), strontium (5 ppm), and cerium (6 ppm) [10]. Thus, the application of such quartzite pebbles should only contribute silica to the glass.

When considering the second silica source, sand, things become a bit more complicated. In general, sands used as silica sources are thought to have also provided considerable amounts of alumina, as well as iron oxide, lime, and magnesia [4, 11]. However, sands are not uniform and can have a wide range of compositions depending on the local geology, as well as the degree to which they are subjected to weathering and alteration [10]. For example, sands from limestone areas, such as those near Amarna in Egypt, can have high lime contents (Table 3.1) [4, 10, 11]. In contrast, sands from more mixed sandstone/limestone areas have lower lime content and increased content of iron and alumina [10]. What is perhaps surprising is the purity of sands from large Aeolian systems such as the Great Sand Sea, which provide widely available sources of high quality sand in Egypt. As a consequence, it has been pointed out that although quartzite pebbles are normally suggested as the silica source of early Egyptian glass, the apparent wide availability of fairly pure sands provides the possibility that such sands were also used [10]. The support for the use of quartzite pebbles in these early glasses stems from the fact that analysis of these early glasses reveals less than 1% alumina [6], which is inconsistent with the majority of sands analyzed.

3.2 Alkali Sources

The two primary sources of alkali for early glassmaking were *natron*, a naturally occurring mineral source, and various types of plant ash [4, 7, 10–13]. *Natron* is a naturally occurring evaporite that forms in evaporitic lakes of Egypt and Syria, most notably the Wadi Natrun, approximately 100 km west of Cairo [4, 10–13]. Here, a group of lakes rises up annually in this large, flat basin due to flooding of the Nile, the lakes apparently fed by underground seepage. As the hot desert sun evaporates the resulting lakes, *natron* precipitates along with other salts [12].

Strictly speaking, *natron* in its modern use is the mineral name for $Na_2CO_3 \cdot 10H_2O$. However, in older sources, this term is also used in a more general sense to mean the deposits of the Wadi Natrun, which can include a variety of evaporite minerals, as shown in Table 3.2 [14]. In this respect, the term *natron* appears to apply to a mixture of evaporites in which the mineral *natron* is limited and the most common form of carbonate is the mineral *trona*, the sodium sesquicarbonate $Na_2CO_3 \cdot NaHCO_3 \cdot 2H_2O$ [14, 15]. It is this later use of the word that accurately describes the *natron* applied as an alkali source for the production of glass. *Natron* as a mixture of evaporites is still a generally pure sodium source, being relatively free of potassium and magnesium, and glasses made with *natron* usually contain less than 1% MgO and less than 1% K_2O [10]. The actual composition of *natron*, however, often varied widely depending on the exact mixture of minerals collected, as well as the inclusion of various other impurities, particularly sodium chloride, sodium sulphate, and silica [11, 14, 16]. The analysis of an ancient *natron* mixture dating to the fourteenth century is given in Table 3.3.

As one might imagine, the composition of plant ash could be even more complex and variable than that of *natron*. In addition to providing sodium and potassium carbonates, such ashes often furnished sodium and potassium salts of both chloride and sulfate, as well as calcium and magnesium salts of carbonate and phosphate [11]. The exact chemical composition of these ashes could also be quite variable due to the fact that the mineral content of the ash depended largely on the soil in which the plants grew. This is best illustrated by the fact that plants grown in salty soil or near the sea produced ash high in soda, while those grown inland gave ash with higher potash content [2]. Overall, plant ash usually has higher amounts of magnesia and potash than does *natron*, and thus glasses produced from these alkali sources generally exhibited higher concentrations of potassium oxide (K_2O, 1–4%) and magnesium oxide (MgO, 3–7%) [7, 10].

The analysis of Mesopotamian tablets provides the word '*ahussu*' as a material used by the Assyrians in glassmaking, which has been interpreted as a plant ash [3, 7]. This ash is specified as coming from the '*naga*' plant [12], which is thought to be almost certainly a desert plant, possibly *salsola kali*. The use of this material suggests that Mesopotamian glassmaking most probably used an alkali derived from plant ash [3, 7]. In fact, it is believed that the glassmakers of Mesopotamia and Persia usually favored various types of plant ash as the alkali source, while *natron* was favored on the Eastern Mediterranean as it was readily available from

Table 3.2 Evaporite minerals from the Wadi Natrun[a]

Mineral name[b]	Chemical formula
Natron	$Na_2CO_3 \cdot 10H_2O$
Thermonatrite	$Na_2CO_3 \cdot H_2O$
Trona	$Na_2CO_3 \cdot NaHCO_3 \cdot 2H_2O$
Nahcolite	$NaHCO_3$
Pirssonite	$Na_2CO_3 \cdot CaCO_3 \cdot 2H_2O$
Burkeite	$Na_2CO_3 \cdot 2Na_2SO_4$
Thenardite	Na_2SO_4
Mirabilite	$Na_2SO_4 \cdot 10H_2O$
Halite	$NaCl$

[a] Ref. [14]
[b] Nomenclature follows International Mineralogical Association recommendations

Table 3.3 Analysis of an ancient *natron* mixture (ca. fourteenth century BCE)[a]

Component	Chemical formula	% by Weight
Sodium chloride	$NaCl$	30.6
Sodium sulfate	Na_2SO_4	20.6
Silica	SiO_2	10.0
Sodium bicarbonate	$NaHCO_3$	12.6
Sodium carbonate	Na_2CO_3	4.9
Calcium carbonate	$CaCO_3$	2.0
Magnesium carbonate	$MgCO_3$	1.9
Alumina	Al_2O_3	0.7
Iron oxide	Fe_2O_3	0.3
Water	H_2O	4.7
Organic matter[b]	Unknown[c]	1.7

[a] Ref. [16]
[b] By difference
[c] The organic material thought to consist of impurities together with herbs and flowers, etc., introduced into the natron mixture for their fragrance

northern Egypt [2, 3]. It has also been reported that lead salts were sometimes used in Mesopotamia as an alternative to the more common sodium fluxes in glass production [17].

While both plant ash and *natron* were used in this initial period of glass production, glass throughout the Eastern Mediterranean, Egypt and Mesopotamia was characterized by high magnesia and potash content [3, 4, 9]. Many authors have linked this increased content with the nature of the alkali used in the glass batch, resulting in the belief that the use of plant ash predominated this early time period [3, 4]. Thus, for the period from 1500 BCE to about 800 BCE, it is commonly held that the majority of glass in both the Near East and Egypt was produced from ground quartz pebbles and the ash from halophytic desert plants [6]. In addition to the necessary soda flux, the plant ash provided calcium and magnesium content

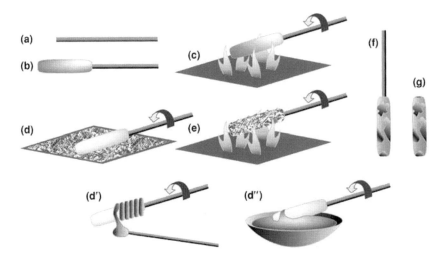

Fig. 3.1 Core molding for the production of hollow glass vessels: initial metal or wooden rod (**a**); formation of core form onto the rod (**b**); firing to set the core (**c**); application of glass via rolling in crushed glass (**d**); firing of applied layer (**e**); completed object (**f**); finalized vessel after removal of rod and core (**g**). Alternate methods for the application of glass included coiling strands of softened glass around the mold (**d′**) or dipping the core in molten glass (**d″**)

which would act as chemical stabilizers for the resulting glass [6]. However, it has also be reported that Mesopotamian tablets have mentioned the use of sea shells and calcinated corals as reagents for glass production, both of which could also have acted as sources of calcium for glass stabilization [7, 12].

3.3 Core-Molding

The earliest method for the production of hollow glass vessels is described as *core-molding* or *core-forming*. This technique dates to ∼1500 BCE [7, 9, 18, 19] and is thought to originate from the Mitannian or Hurrian regions of Mesopotamia, after which it very quickly spread to Egypt [6, 20]. As outlined in Fig. 3.1, this method was accomplished by first shaping a form or core for the vessel onto the end of a wooden or metal rod (Fig. 3.1a and b) [4, 18, 20]. The primary requirement of the core material was that it needed to be pliable enough to form, solid enough to hold a given shape, yet should be weak enough to easily crumble for later removal from the finished vessel [21]. The development of such a core material would have been difficult, as it had to possess a number of contradictory properties, including the need to allow the glass enough surface purchase to adhere during the forming process, yet remain separate enough to allow easy removal on cooling [21]. The precise composition of the core has generated much speculation and a number of different materials have been proposed, including clay or mud [3, 4, 12, 21], sand

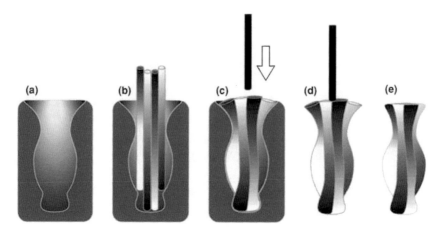

Fig. 3.2 Fuse-casting of glass objects: production of black mold (**a**); glass pieces added, heated to fuse and fill mold (**b**); metal rod inserted (**c**); mold removed (**d**); piece ground and polished to finish (**e**)

[3, 4, 12, 18, 19], fabric over a base material [12, 21], sand-lime mixtures [16], mixtures of clay and horse dung [22], and even camel dung [21]. The formed core could then be heated or fired in order to help set its shape (Fig. 3.1c).

The fabrication of the vessel itself would then begin by building up layers of glass around the central set core. This was accomplished by a number of different methods, including treating the core with an organic binder (egg white or honey) and rolling in crushed glass (Fig. 3.1d) [4, 21], winding strands of hot, softened glass around the core (Fig. 3.1d′) [3, 20, 21], or repeatedly immersing the core in molten glass not much above the softening temperature [3, 12, 18, 20–23]. This would then be heated to give a uniform layer of glass (Fig. 3.1e), cooled, and a new layer applied. In such a repetitive manner, the wall of glass would be built up iteratively until the desired thickness had been achieved [18, 20]. While the wall of glass was still soft, the vessel could be rolled on a slab of stone in order to smooth the surface or further decorated via various manipulations [4, 12, 21]. Once the final workings were completed and the object had cooled (Fig. 3.1f), the rod was removed from the core. The core material was then carefully dug from its center to give the finished hollow vessel (Fig. 3.1g). As one might imagine, this method is limited in scale and the inside surfaces of the finished vessel were extremely contaminated through contact with the core [4, 21]. Generally, the hollow interiors were not of significant volume, nor were vessels produced by this method very large in size—probably five inches or less in height [18].

3.4 Cast Glass

The forming of glass then made another significant advancement in about 1200 BCE, when the Egyptians learned how to press glass into open molds [18, 24, 25]. This now made possible the production of bowls, dishes, and cups which could not be made by the previous core molding methods. Casting involved melting glass (ingots, pieces, rods, or grains) into a mold which provided the shape of the desired object (Fig. 3.2) [26, 27]. The mold could then be broken away from the glass after it had cooled [26]. This method seems to have been used for the production of relatively simple shapes of limited sized and complexity [27].

In order to make the glass fill the mold, the temperature would need to be raised and kept at a relatively high level. At such temperatures the glass would lose its entire original surface and take on the texture of the mold material. Such mold materials would have also been relatively simple and would have contaminated the glass surface [27]. In addition, it is very difficult to obtain objects free from trapped air bubbles and unsightly portions of crystallized material representing the original exposed surfaces of the various glass pieces with which the mold was filled [26]. For these various reasons, it would have been necessary for the object to be ground and polished to remove such contamination and defects. Unfortunately, the fact that all objects made by casting have been completely ground and polished has also removed all traces that could act as clues to the exact method used for their production [27].

In order to limit some of the defects discussed above, glass could be melted separately and then poured into the mold from a pot or ladle, but even then it is very difficult to prevent the inclusion of bubbles. In addition, the molten glass never attains the mobility of molten metal [26, 27]. Thus, the casting of glass is a much slower process and involves the gradual feeding of glass into the mold while subjecting it to continuous heating over a long period [27]. The lower mobility of glass also makes it so that it will not flow through small orifices and will not properly fill molds with intricate patterns, thus limiting this method to the production of larger objects of simple design [26].

3.5 Decline and Renewal

After this period, the glass industry declined and there is little evidence of further evolution until glassmaking revived in Mesopotamia about 700 BCE, followed by a similar revival in Egypt about 500 BCE [1, 8, 12]. This resurgence of glass was part of the Iron Age revival of culture that followed the period of turmoil in the Mediterranean between 1200 and 1000 BCE. After this initial revival, production spread to new glassmaking centers, which introduced variations of established processes, and in some cases, developed new techniques [1]. Thus, for the next 500 years, glassmaking centers developed in Egypt, Syria, and the other countries along the eastern shore of the Mediterranean Sea [8], with the industry gradually becoming centralized at Alexandria [20].

Glass during the period of sixth century BCE to about fourth century CE is distinguished by its antimony-rich nature [17]. This glass is characterized by a lower potassium (0.1–1.0%) and magnesium (0.5–1.5%) content [3, 9], along with a consistent appearance of antimony in high concentration. This glass represents the composition used in Greece, Asia Minor and Persia during the fifth and fourth centuries BCE and continued to be popular in areas from the Euphrates eastward during the ascendancy of Rome [9]. Many authors have linked this variation in the magnesia and potash with a change in the nature of the alkali used for the production of glass, proposing a move from the earlier application of plant ash to the use of *natron* [3, 4].

Antimony was known in Mesopotamia quite early and was of crucial import for the manufacture of opaque glasses [7]. However, at this point, a major change in glass manufacture was a shift in emphasis from opaque to clear glass. This move to clear and translucent colored glass had as much to do with a shift in viewpoint as with any improvement in technology [1]. With this shift in emphasis, antimony became a commonly applied early decolorant for the production of colorless glass [4, 17]. The source of the antimony was probably stibnite (Sb_2S_3) which can be found throughout the Mediterranean and the Near East [28]. Prior to its addition to the glass batch, the stibnite raw material would first be roasted. The roasting process would initially convert the sulfide to the corresponding oxide (Sb_2O_3). This initial oxide was then converted to the mixed valence product Sb_2O_4 with further heating at temperatures of 460–540°C in dry air. When added to the glass, Sb_2O_4 acts as an oxidizing agent to convert the strongly absorbing Fe(II) species to the fairly colorless Fe(III), thus removing the coloring due to the iron impurities [17]. The production of clear, antimony-rich glasses dates back to about the seventh century BCE [12].

3.6 Roman Glass

During the fourth century BCE, Greek culture spread over the Near East due to the wide-spread conquests of Alexander the Great (d. 323 BCE) and became the dominant cultural element in Syria, Egypt, Asia Minor, Mesopotamia, Southern Italy and Sicily (Fig. 3.3) [29]. Because of this, the Greeks became exposed to the technological knowledge of the middle-east (Mesopotamians, Babylonians, Syrians, Persians), as well as that of the Egyptians, Indians, and Chinese. Rome conquered Greece in ∼200 BCE and by the first century CE had engulfed the entirety of the Mediterranean. In the process, the Romans absorbed Greek culture and natural philosophy, including the collected knowledge and technology of glassmaking.

Glass of the period fourth century BCE to ninth century CE represents the normal composition of "Roman glass" and was probably typical of Syrian coastal cities, Egypt, Italy, and the western provinces. The composition is similar to that of the antimony-rich group except that the antimony content is much lower and in most cases the manganese content is much higher. This suggests that the distinction between the Roman glass and the previous antimony-rich glass simply represents a change in the decolorant used [9].

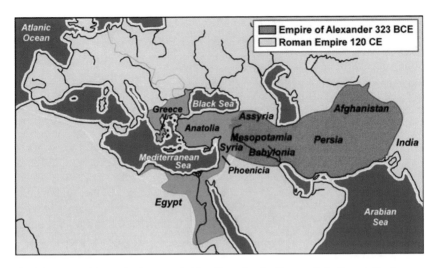

Fig. 3.3 Maps of the territories occupied by Alexander the Great and the Roman Empire

It is believed that the Roman period also marked another change in the type and origin of the raw materials used to produce glass [6]. *Natron* continued to be the primary source of alkali, most probably from the Wadi Natrun in Egypt [3, 4, 6, 28]. The Romans extensively exported *natron* and it remained the alkali of choice for glass production until the decline of the Roman Empire [3]. However, instead of the ground quartz pebbles thought to be favored in the previous periods of glass production, the increased alumina content of the Roman glass suggests a move to sand as the silica source [6]. Due to the higher calcium content of such siliceous–calcareous sands (Table 3.1), these acted as a source of both silica and the lime [6, 30]. The use of sand in Roman glass is confirmed by Pliny the Elder in his description of first century glass production [31]:

> ... at the present day, there is found a very white sand for the purpose, at the mouth of the river Volturnus, in Italy. It spreads over an extent of six miles, upon the sea-shore that lies between Cumæ and Liternum, and is prepared for use by pounding it with a pestle and mortar; which done, it is mixed with three parts of nitre, either by weight or measure, and, when fused, is transferred to another furnace. Here it forms a mass of what is called "hammonitrum;" which is again submitted to fusion, and becomes a mass of pure, white, glass. Indeed, at the present day, throughout the Gallic and Spanish provinces even, we find sand subjected to a similar process.

This combination of sand and *natron* for glass production is thought to have been first introduced somewhere in the eastern Mediterranean during the eighth–seventh centuries BCE and continued throughout the Roman and Byzantine periods until about 850 CE [6].

As mentioned above, the other primary shift in raw materials for Roman glass production was the move from antimony to manganese as the decolorant [4, 9, 17]. The production of clear glasses containing manganese dates back to the first century BCE [12]. While antimony is considered a better decolorant than manganese, the Romans probably preferred the latter due to its availability [17]. In all regions with strong Roman influence, the level of antimony in glass decreased, while manganese increased. This change is considered another important chemical fingerprint for glass technology [17]. The source of manganese was manganese dioxide (MnO_2), in which the Mn(IV) acts as an oxidizing agent. Thus reaction of Mn(IV) with the strongly absorbing Fe(II) species produces Fe(III) and Mn(II), both of which are essentially colorless [4].

Other materials used in the production of Roman glass include lead salts and possibly various species that would have acted as sources of lime [12, 17, 28, 31]. Some have claimed that the uniform calcium content of Roman glasses is evidence of the intentional use of lime in glass by the Romans [12]. Chalk, limestone or burned shells have all been suggested as convenient sources of lime that could have been raw materials for Roman Glass [12] and Pliny the Elder does mention in passing the addition of shells and fossil sand to glass [31]. Others, however, have pointed out that the distinct lack of significant amounts of such components in known glass recipes does not support this. As these components are also not mentioned in the known glassmaking treatise of the medieval and Renaissance periods, it is currently held by most that the importance of lime was not recognized at this point, nor was it intentionally added as a major constituent before the end of the seventeenth century [32]. In some cases, it was known that lead salts would be intentionally added to glasses and enamels to improve the working properties of the melt [28] or as a means to enhance the brightness of the glass [17].

In general, the uniformity in its full composition (not just the calcium content as discussed above) is a particular characteristic of glass found throughout the Roman Empire. Some have proposed that this suggests that a limited number of sites were used for the production of glass and that glass cullet from these production centers was transported around the Empire for working into glass vessels at a significantly larger number of local facilities [6]. This proposed production model is supported by both the scale of the tank furnaces used for glass production, and the fact that the only large-scale source of *natron* is the Wadi Natrun in Egypt. Any other production model would require the transport of large quantities of *natron* around the Empire, which is significantly more bulky in volume and thus more difficult [6].

Most of the currently investigated centers for the production of glass are located in the Levant and Egypt, dating to the late Roman to early Byzantine periods. The glass products of these manufacturing centers can be distinguished from each other on the basis of chemical composition. For example, the relative amounts of alumina, lime, magnesia and iron oxide in the glass reflect the compositions of the different quartz sands used at the different production centers. While this is supportive of the limited number of production sites theorized above, the glass from the currently investigated sites does not match the composition of earlier Roman glass dating from first to third centuries CE. Therefore, it remains uncertain whether this production model is reflective of the Roman Empire in general or just limited to the later Roman period [6].

3.7 Slumped Objects

Roman civilization provided a ready market for high quality secular glass items. This in turn encouraged the development of new processes and a more centralized approach to glassmaking. For the first time the production of large numbers of similar items became an economic aim and genuinely industrial systems were created. This began with the process of bending, also called sagging or slumping (Fig. 3.4) [27]. Here, hot glass was poured onto a flat surface (Fig. 3.4a) and then pressed with a flat, disk-shaped former (Fig. 3.4b). Once cooled, the former was removed to create a glass disk (Fig. 3.4c), which was then transferred onto a "former" mold of fired clay or cut stone (Fig. 3.4d) and heated to soften the glass disk. The combination of heat and gravity would cause the disk to sag over the mold to give a bowl-shaped glass object (Fig. 3.4e). This technique was probably derived from the similar forming of a metal blank over a mandrel by hammering. The slumped glass bowls made by these methods were still individually made and would still have been exclusive items, produced for the social elite. By 400 BCE, large scale production of slumped objects was in evidence [27].

The glass bowls were finished by grinding and polishing inside, to remove mold markings, and were often completed by the addition of a single horizontal ground band. In this and their general profiles, they copied analogous metal objects— another clue as to their origins. However, the material of glass transformed the original metal species into the unique single-color, refined vessels that characterize the process [27].

3.8 Glassblowing

A major new development during the Roman period was the introduction of glassblowing during the first century BCE. Glassblowing is thought have been discovered during the period between 300 and 20 BCE [2, 6, 12, 18, 20, 24, 25] and its origin and development is typically attributed to Phoenician craftsmen, who were among the first in Syria to produce blown glassware [1, 2, 18, 20]. The art of glassblowing is thought to have then come to Rome with craftsmen and slaves after the area's annexation by the Roman Empire in 63 BCE [1, 4].

It has been proposed that this was possibly an accidental discovery resulting from the production of glass tubes. It has been suggested that while an artisan was working with such a glass tube (perhaps cutting it into beads), the end of the heated tube collapsed. Thinking to open it by force of air, the artisan could have blown into the tube, which instead of opening, formed a bubble at its end, which expanded as he continued to blow. Of course, we have no evidence to support this theory, but the idea seems plausible [18].

Technically, this process required heating glass to a much higher temperature than for either casting or molding and needed the glass to be very fluid. Laboratory experiments reconstructing the conditions of ancient glassmaking have shown that in

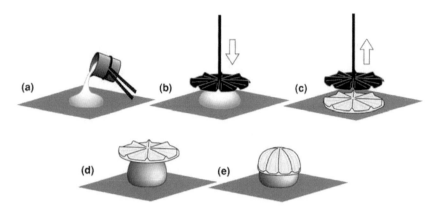

Fig. 3.4 The formation of open-form bowls by sagging glass over convex "former" molds: molten glass is poured onto a flat surface (**a**); pressed with a flat, disk-shaped former (**b**); cooled to create a glass disk (**c**); transferred onto a "former" mold (**d**); heated to cause the disk to sag over the mold giving the final bowl shape (**e**)

order to make the typical Roman glass fluid enough for glass blowing it had to be heated to at least 1080°C [12]. To achieve such temperatures would have required a knowledge and experience of furnaces that had developed in the glass industries of the Middle East [19]. In terms of the blowpipe, all are agreed that it was about the same size and shape as it is today—a hollow iron tube 4–5 feet long, with a knob at one end and a mouthpiece on the other. This simple device, in the hands of skilled artisans, made possible the creation of an almost endless variety of hollow glass objects [20].

With the introduction of glass-blowing, very thin, transparent glass could be produced and this technique enormously increased the versatility of glass and in particular, opened up potentially new uses [19]. The whole character of glass vessels changed: thin-walled vessels began to replace the heavier forms of earlier periods and the scale of glass production increased dramatically with large tank furnaces (2 × 4 m in size) being used to produce several tons of glass in a single firing [6]. The new technique made possible the rapid production of simple utilitarian vessels in much larger quantities, and glass became a household commodity [6, 12]. A middle-class Roman family probably owned glass storage containers, drank from glass vessels, and bought souvenir glass cups with the names of favorite gladiators molded into them. Even the final resting place of many Romans was a glass funerary urn [12].

3.9 Windows

Another significant innovation introduced by the Romans was the application of glass to the construction of window panes [33]. Such window glass was produced by the Romans as early as the first century CE, and is commonly found in Roman sites in Britain [34]. Samples have also been uncovered at Pompeii (destroyed in 79 CE),

Fig. 3.5 Modern
reproduction of Roman
window glass ($\sim$ 5 mm thick)
[Copyright *Roman
Glassmakers*, used with
permission]

some as large as 30 $\times$ 40 inches [4]. Such early windows were usually quite small, of irregular thickness, and not truly clear and transparent (Fig. 3.5) [33].

Early window panes were produced using one of several different processes [33–36]. The earliest method is known as "cast glass", and panes produced by this method were of uneven thickness, with one side exhibiting a fire polished or "glossy" texture and the other side a pitted, matt finish. The Romans ceased to use cast glass in the third century CE and thus the precise technique of making cast glass has been lost [34, 35]. It has often been suggested that such panes were produced via the pouring of molten glass into a mold in much the same way metals are cast. However, arguments against this method are that it would not result in the forms of the edges and corners, nor the tool marks seen on original Roman glass [34, 35].

Another method often suggested is casting of soft, hot glass on a flat surface and then pressing it into some semblance of flatness with a moist, wooden mallet [33]. Contemporary glassmakers have been able to produce glass panes in this manner that closely match those seen on original Roman panes [34, 35]. In this process, molten glass was poured onto a damp surface and immediately flattened with a large block of damp wood to produce a flat disc of glass about 5 mm thick. The hot disc was then pulled and stretched in order to form a rectangular shape. The tools used were all simple metal rods, hooks and pincers, and the marks produced on the finished pane closely match those seen on original Roman panes [34, 35]. While crude, this method is straightforward and repeatable and it is easy to imagine its use during the early Roman period.

Another method used by the Romans for the production of window glass was called the "crown method". This process consisted of first blowing a hollow sphere on the end of a blowpipe, after which the end opposite the blowpipe was opened. The opened, soft sphere was then vigorously rotated such that centrifugal force would cause the glass to flatten out into a disk. The disc was then cooled and cut into small sheets. Every disk has a lump of glass at the center (known as the "bull's eye" or "crown"), and only small sheets could be made by this process [36].

As can be seen from the discussion above, glass manufacturing developed to a new height during the Roman Empire and spread from Italy to all of the Roman provinces [8, 12]. Class objects from the Roman period are found in abundance out to the ends of the Empire, as far north and west as Scandinavia and Britain and as far east and south as Syria and Ethiopia. The Romans are usually praised primarily for their practical skills rather than for their aesthetic achievements, but in the case of glass they excelled on both counts [12]. For this reason, the first four centuries of the Common Era are often referred to as the *First Golden Age of Glass* [8].

References

1. Cummings K (2002) A history of glassforming. A & C Black, London, pp 102–133
2. Kurkjian CR, Prindle WR (1998) Perspectives on the history of glass composition. J Am Ceram Soc 81:795–813
3. Shortland AJ, Tite MS (2000) Raw materials of glass from Amara and implications for the origins of Egyptian glass. Archaeometry 42:141–151
4. Lambert JB (1997) Traces of the past. Unraveling the secrets of archaeology through chemistry. Addison Wesley, Reading, p 109–116
5. Smirniou M, Rehren TH (2011) Direct evidence of primary glass production in late bronze age amarna Egypt. Archaeometry 53:58–80
6. Tite MS (2004) Glass and related vitreous materials. Proc Intern School Phys Enrico Fermi 154:369–376
7. Saldern AV, Oppenheim AL, Brill RH, Barag D (1988) Glass and glassmaking in ancient mesopotamia. Corning Museum of Glass Press, Corning, pp 21, 67, 79, 109–111, 184
8. Axinte E (2011) Glass as engineering materials: a review. Mater Des 32:1717–1732
9. Sayre EV, Smith RW (1961) Compositional categories of ancient glass. Science 133:1824–1826
10. Shortland AJ, Eremin K (2006) The analysis of second millennium glass from Egypt and mesopotamia, Part 1: new WDS analyses. Archaeometry 48:581–603
11. Turner WES (1956) Studies in ancient glasses and glassmaking processes. Part V. Raw materials and melting processes. J Soc Glass Technol 40:277T–300T
12. Brill RH (1963) Ancient glass. Sci Am 109(5):120–130
13. Moretti C (1985) Glass of the past. Chemtech 15(6):340–344
14. Shortland AJ (2004) Evaporites of the Wadi Natrun: seasonal and annual variation and its implication for ancient exploitation. Archaeometry 46:497–516
15. Shortland AJ, Degryse P, Walton M, Geer M, Lauwers V, Salou L (2011) The evaporitic deposits of Lake Fazda (Wadi Natrun, Egypt) and their use in roman glass production. Archaeometry 53:916–929

16. Noble JV (1969) The technique of Egyptian Faience. Amer J Achaeol 73:435–439
17. Lambert JB (2005) The 2004 Edelstein award address, the deep history of chemistry. Bull Hist Chem 30:1–9
18. Roger F, Beard A (1948) 5,000 years of glass. J. B. Lippincott Co., New York, pp 23–25
19. Macfarlane A, Martin G (2002) Glass a world history. University of Chicago Press, Chicago, pp 10–26
20. Philips CJ (1941) Glass: the miracle maker. Pitman Publishing Corporation, New York, pp 6–10
21. Cummings K (2002) A history of glassforming. A & C Black, London, pp 26–27 48–55
22. Tait H (ed) (1991) Glass, 5,000 years. Harry N. Abrams, Inc., New York, p 214
23. Morey GW (1954) The properties of glass. 2nd edn. ACS monograph series no. 124, Reinhold Publishing, New York, pp 8–10
24. Holmyard EJ (1990) Alchemy. Dover Publications, New York, pp 47–49
25. Talyor FS (1992) The alchemists. Barnes & Noble, New York, pp 39–46
26. Philips CJ (1941) Glass: the miracle maker. Pitman Publishing Corporation, New York, pp 173–174
27. Cummings K (2002) A history of glassforming. A & C Black, London, pp 60–64,103
28. Mass JL, Wypyski MT, Stone RE (2002) Malkata and Lisht glassmaking technologies: towards a specific link between second millennium BC metallurgists and glassmakers. Archaeometry 44:67–82
29. Clagett M (1994) Greek science in antiquity. Barnes & Noble, New York, p 22
30. Silvestri A, Molin G, Salviulo G, Schievenin R (2006) Sand for roman glass production: an experimental and philogical study on source of supply. Archaeometry 48:415–432
31. Pliny the Elder (1855) The natural history. Bostock J, Riley HT (trans) Taylor and Francis, London, Book XXXVI, Chapter 66
32. Turner WES (1956) Studies in ancient glasses and glassmaking processes. Part III. The chronology of the glassmaking constituents. J Soc Glass Technol 40:39T–52T
33. Roger F, Beard A (1948) 5,000 years of glass. J. B. Lippincott Co., New York, pp 137–138
34. Taylor M, Hill D (2001) No pane, no gain. Glass News 9:6
35. Taylor M, Hill D (2002) An experiment in the manufacture of roman window glass. ARA Bull 13:19
36. Philips CJ (1941) Glass: the miracle maker. Pitman Publishing Corporation, New York, p 199

Chapter 4
Reinventing an Old Material: Venice and the New Glass

In the fourth century CE, the stability of the Roman Empire began to rapidly diminish. As the central cohesion of the Empire was lost, glassmaking centers began to more greatly reflect their regional influences and many of the more sophisticated techniques became less widespread [1]. Shortly before 300 CE, Diocletian became Emperor and tried to stabilize the Empire by dividing it into Western and Eastern halves (Fig. 4.1). Each half of the former Empire had its own capital (Rome in West and Constantinople in the East) and Emperor, although the Emperor of the West was subservient to that of the East [2, 3]. The fragmentation of the empire, first into East and West, and later into isolated regions conquered by outside forces, meant the end of centralized glass production. Glassmaking shifted from urban centers to rural locations closer to sources of fuel [1]. As a result, glassmakers within the two halves of the empire became isolated and eastern and western glassware gradually acquired distinct characteristics. Still, glassmaking was able to survive the end of the unified Roman system and adapted to the needs of the new political framework. The primary result of this changing framework was the loss of more specialized and sophisticated decoration techniques, such as cutting, polishing, and enameling. Critical techniques such as glassblowing were simplified to their basic essentials and simple procedures like mold blowing essentially disappeared [1].

4.1 Glass in the West

Glasshouses had been established under Roman rule in the western provinces of Gaul and Brittany during the first through third century CE. These included sites as far north as Boulogne, Trier, and Cologne, as well as the British sites of Manchester and Leicester [4–6]. Here, production was centered in forested areas that could provide plentiful fuel for the furnaces, resulting in the establishment of

S. C. Rasmussen, *How Glass Changed the World*,
SpringerBriefs in History of Chemistry, DOI: 10.1007/978-3-642-28183-9_4,
© The Author(s) 2012

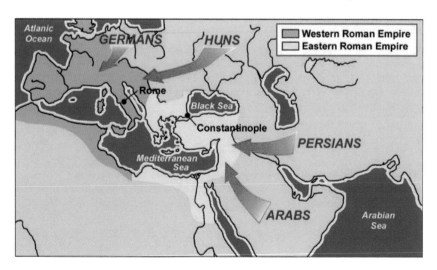

Fig. 4.1 The Roman Empire ca. 350 CE

glassmaking centers in the forests of Northern and Central Europe [1, 7]. By ~ 500 CE, the Western Empire fell to German tribes and glassmaking essentially ceased in the West for a time. The established glasshouses survived in the forested regions, however, and the knowledge of glass production was not completely lost. Glass was produced locally, reflecting local needs, raw materials, and glassmaking systems. Reduced access to raw materials eventually made it impossible to achieve colorless glass by careful selection of raw materials and glass inevitably showed the character of the local silica and flux used [1]. Such northern glass produced in the Middle Ages was sometimes referred to as "waldglas" (forest glass), and was commonly dark green or brown in color as a result of the impurities present [8].

One important raw material in decreased supply was soda imported from Egypt, critical for the production of high-quality soda-lime glass [9]. This limited supply may have been due to either the absence of reliable trade routes or an increase in the cost of soda to the point that is was no longer cost-effective to import [6]. Nevertheless, in response to this lack of available soda, several northern glasshouses started to use the ash of wood logs as the primary flux for glass production as early as 800 CE [7, 9]. The tree commonly used for this purpose was beech, which produced an ash containing significant amounts of calcium, potassium and magnesium oxide (Table 4.1). Oak was also used, but not as commonly as beech [10].

It is important to remember that the mineral content of plant ash depends largely on the soil in which the plants grow, as discussed in the previous chapter. As such, ash from plants grown in salty soil or near the sea are typically high in soda, while ash from inland grown species are typically higher in potash content [8]. Thus, as shown in Table 4.1, the ash of the various inland grown trees are very low in soda, but all exhibit significant potash content (up to 37%). Other than mulberry, the preferred beech ash gave the highest potash content [11]. However,

Table 4.1 Composition of ash from various tree sources[a]

Ash source	% Composition of ash[b]				
	SiO_2	CaO	MgO	Na_2O	K_2O
Oak	2.0	72.5	3.9	3.9	9.5
Apple	2.7	70.9	5.5	1.9	11.8
Mulberry	3.6	57.0	5.8	6.6	36.6
Beech, trunk	5.4	56.4	10.9	3.6	16.4
Beech, brushwood	9.8	48.0	10.6	2.4	13.8
Beech, leaves	33.8	44.9	5.9	0.7	5.2

[a] Ref. [11].
[b] % by weight

as with all of the sources shown in Table 4.1, it also gave very high levels of lime, with the average CaO/K_2O ratio in the ash from beech trunks ranging from about one to two. The portion of the beech tree used could also significantly affect the chemical composition of the resulting ash and CaO/K_2O ratio in the final beech ash depended on the proportion of trunks to twigs or branches used in the generation of the ash [7].

Thus by the tenth century CE, glass in the northern glasshouses was produced from a combination of the tree ash and local sands to give a potash-lime formula [6, 9, 10, 12]. Chemical analysis of northern glasses of this time period have revealed high potassium and calcium content (11.8 and 17.9%, respectively) and low sodium content (1.63%). However, the potash-lime glass produced from 780 to 1000 CE is not as chemically well defined as the later specimens, with CaO/K_2O ratios ranging from as low as one to as high as six [7].

The chemical differences between the potash-lime and soda-lime glasses resulted in very different physical properties. For example, the use of soda as a flux could reduce the melting point of the silica to below 1000°C, while the similar application of potash resulted in melting as low as 750°C [13, 14]. In addition, potash-lime glass was typically harder than soda-lime glasses, making it better for cutting and engraving, although it was also heavier and often not as clear [15]. Due to its lower melting temperature and the simple availability of the oak and beech in the Northern European forests, potash-lime glass could be readily and inexpensively mass-produced, making it very desirable, particularly for the production of windows [7, 10].

4.2 Glass in the East

The development of glass in the East took an entirely different direction despite sharing the same initial origins. As with the Western Empire, the East was also under threat by exterior forces. However, as it was wealthier and more populated than the west, it thus had greater resources to withstand the initial Germanic invasions, as well as Persian forces from the east. While the emperor Heraclius successfully defeated the Persians in 627 CE, this effort had significantly weakened the Eastern Empire. From this point on, the East was no longer referred to as the Eastern Roman Empire and became known as the Byzantine Empire (although Byzantine emperors continued to call themselves Roman) [3].

Fortunately glassmaking continued in the Byzantine Empire long enough to ensure its survival, and aspects of glassmaking that died out in the West were thus kept alive in the East. Its capital Constantinople was built on the site of the ancient city of Byzantium and its eastern cultural influences caused glassmaking to develop in a very different way than the western half of the old Roman Empire. Although not enough glass from the early Byzantine period has survived to recreate its exact development, there remains enough evidence to suggest that important aspects of working, using and valuing glass flourished. The few surviving glass artifacts demonstrate both the ability to make clear glass and the application of decorative techniques such as cutting, gilding, and enameling [1]. In its continued development, Byzantine glassmakers brought to glass a highly developed artistic taste, particularly in the use of color [4]. Here glass was used for the production of high status items, such as jewelry, mosaics that utilized hundreds of thousands of tiny glass tesserae, and even chalices for the Christian mass [1, 5].

In its weakened state following the conflict with the Persians, the Byzantine Empire was then attacked by Muslim Arabs in ∼635 CE and by ∼650 CE all that remained of the Eastern Empire was Asia Minor, the Balkans, North Africa, Sicily, and Italy [3, 16]. The Empire was even further reduced, losing North Africa and parts of Italy, until ∼673 CE, when the Byzantines developed *Greek fire*[1] which was used to drive off the Arab fleet [17]. Their fleet destroyed, the Muslims finally made peace.

The East was now divided into the Byzantine Empire and a growing region dominated by the rise of Islam. In ∼630 CE, the Muslim Arabs had begun the creation of a new empire and civilization that was eventually comparable to that of the height of the Roman Empire. This new empire generated centralized, sophisticated urban order, and in doing so ensured the survival of types of glassmaking and decoration that would have otherwise disappeared entirely. Contact between the Byzantine Empire and the new empire of Islam allowed Islamic glassmakers to add the known Roman and Byzantine glassmaking techniques to their own glassmaking knowledge [1]. As with many chemical arts, this cumulative glassmaking knowledge was then preserved by the world of Islam until the coming of the Renaissance in the West. In Islam glassmaking flowered again for a time, combining Roman knowledge with indigenous traditions [5].

It was here that a glass composition that has been referred to as *Islamic soda-lime glass* was introduced in the eighth to tenth centuries CE. This glass exhibits a

---

[1] The ingredients for Greek fire were kept a state secret and the precise composition is still unknown. It is thought, however, to be a mixture of naphtha and sulfur, with a mixture of resins or pitch to thicken it. This mixture was then forced through a siphon and ignited by a flame burning at the tip. The resulting spray of flaming jelly then stuck to and burned everything it hit. Supposedly, the only thing that could extinguish it was vinegar, sand, or urine [17].

return to the higher magnesium and potassium concentrations typical of the second millennium BCE, but without showing the low manganese content of these earlier glasses [12]. As with the previous glass of 1500–800 BCE, it is believed that these increased levels of magnesium and potassium are a result of the nature of the alkali used [18, 19] and points to the use of plant ash, rather than the *natron* commonly used throughout the Roman periods. As with the move from *natron* to potash-rich ash in the west, it is believed that this change from *natron* to soda-rich ash was due to increasing difficulties in obtaining *natron* from Eygpt [20].

Due to the close similarity between the Islamic glass and the previous glass of the second millennium BCE, it has been reasoned that Islamic glass represents an uninterrupted continuation of that early glassmaking tradition. However, analyses of glass samples over this time period suggests that a lapse of many centuries occurred between the production of these two categories of glass, with glasses characteristic of both the antimony-rich and Roman glasses being produced in areas destined to become Islamic. Within this same time period, Islamic glass-makers also produced a high-lead glass with considerably more lead and less alkali and lime than the previous lead-containing Roman glasses [12]. After the initial Crusades in the eleventh century, the center of glass manufacture gradually shifted from the Islamic glassmakers to the growing glass industry of Venice and stayed there for at least the next four centuries [4, 5].

4.3 Venice and Murano

It is believed that the original population of Venice consisted of refugees from nearby Roman cities who were fleeing successive waves of Germanic and Hun invasions during the fifth century. These refugees took refuge in the salt marshes of a lagoon and after a while managed to dam back the sea and reclaim from it a small group of islands which became Venice [21]. Although the Western Roman Empire eventually fell to German tribes, the small strip of coast containing Venice was retained by the Eastern Empire and it was eventually recognized as part of the Byzantine territory. Beginning in the ninth century, Venice developed into a city state and grew in importance during the eleventh to thirteenth centuries by exploiting its unique position in the Mediterranean [1]. Its strategic position at the head of the Adriatic and the power wielded by its fleet allowed it to exploit its advantageous trading position, achieving a virtual dominance of trade with the East. As a result, Venice became a flourishing trade center (Fig. 4.2), as well as an important center for learning and the arts. In such an environment, the technology and art of glass flourished from the thirteenth to sixteenth centuries [1, 22], and it was here that significantly improved glass was produced beginning in the second half of the thirteenth century [1, 23, 24].

It seems likely that the tradition of glassmaking never died out in Italy after the fall of Rome, and by the time of the Crusades, glass manufacture had been revived in Venice [22, 26, 27]. The glass industry of Venice was originally developed in

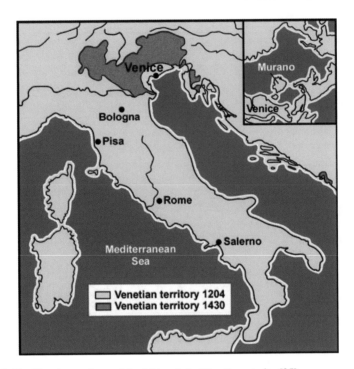

Fig. 4.2 The Venetian territory of the thirteenth to fifteenth centuries [25]

order to produce sheets of richly colored glass for mosaic tesserae. Additional products included simple poured discs made in a wide range of colors, both opaque and transparent, to be traded throughout the world as raw materials to be crushed into powder for enamellers. This simple glass industry was well established by the ninth century and was soon operating on an amazingly large scale [1, 21]. By 1200, the Venetian glass industry was prospering [4].

This initial industry provided a strong foundation on which additional knowledge and technology could be built, particularly techniques and expertise gained from the East. It is believed that the Venetians gained such additional knowledge from Byzantine glassmakers after the sack of Constantinople in 1204 during the fourth Crusade [23]. This influx of Eastern knowledge was then significantly enhanced by a critical treaty signed in 1277 [1, 28] between Jacopo Contartini, the Doge of Venice, and Bohemond VII, Prince of Antioch, to facilitate the transfer of technology between the two centers [1, 26]. This transfer included that of Syrian glassmaking, thus allowing many secrets of glassmaking to be brought to Venice at a crucial point in its development as a world power. Trade documents describe the importation of raw glass (cullet) and plant ashes from the Near East to Venice as early as the thirteenth century [29]. Venice's commercial expansion and extensive shipping network in the eastern Mediterranean secured a continuous supply of plant ash for the Venetian glass industry [24]. The imported Syrian ashes were a cheap commodity.

Better yet, because the ashes were so bulky and heavy, and because Venice needed such large quantities, they were used as ballast[2] to balance the cargos of cotton in the Venetian cogs returning from Syria, and thus were also transported to Venice at low cost [24]. The Venetian cog line to Syria was established in 1366 and operated under strict state control, providing a regular, dependable, and abundant supply of Levantine soda ashes. Annual imports of ash to Venice exceeded 350 t by 1395, with this growing to 1750 t by the end of the fifteenth century [24]. In addition to critical raw materials, Syria also provided physical expertise in the form of Muslim artisans who could directly teach the Syrian techniques to the Venetians [1, 23, 28].

These factors provided key components that led to the flowering of glass in fourteenth to sixteenth century Venice. The combined factors that provided the perfect environment for the dominance of the Venetian glass industry included: (i) the initial existing glassmaking industry; (ii) the influx of knowledge, skill, and materials from Syria; (iii) the growing importance of Venice in terms of trade, commerce and culture; (iv) the wider cultural context of the Italian Renaissance in painting, architecture and the applied arts; and (v) the quality of raw materials available to the Venetian glassmakers. This resulted in a glassmaking center that produced some of the finest glass ever known and dominated glass technology for centuries [1]. As such, the Venetian period is often referred to as the *Second Golden Age of Glass* [27].

As the glass industry grew, the center of Venice became dominated by furnaces blazing night and day. The Venetian glassmakers organized themselves into their own guild in 1268 [1, 23] with a more elaborate guild system established in 1279 [4]. Far too often, control of the intense fires of the glassmaking furnaces was lost, resulting in destruction of not only important glasshouses, but adjacent city neighborhoods as well. As a solution to this danger, the whole industry was ordered to be moved from Venice proper to Murano (Fig. 4.1, inset), a separate borough of the city, in 1291 [4, 23]. Murano was an island about a mile from Venice and already contained a few established glasshouses [21, 26]. Murano gradually became a place of roaring furnaces that devoured wood 24 h a day. The glasshouses are said to have extended for an unbroken mile where thousands of workers toiled at making a variety of glass objects, including windows, vast quantities of beads, bottles, mirrors, and ornamental glassware [4, 21].

By the end of the thirteenth century Venetian glassmaking really began to influence the whole of Europe [22]. The fame of Venetian glass had spread and through the export sale of fine glassware and beads, the Republic grew in wealth and importance. However, Venice knew that the value of their glass would only last as long as other countries could not make competing high quality glass for themselves. As a consequence, in was critical to maintain the secrets of Venetian knowledge and methods. To accomplish this, both penalties and rewards were implemented to maintain secrecy and thus the monopoly of glassmaking for Venice [21].

---

[2] Weight to control the buoyancy and stability of the ship.

The centralization of the glass industry on the island of Murano not only removed the fire hazards from Venice, but also allowed the governing body to better protect the secrets of the Venetian glassmaking methods [21]. The emigration of Muranese glassmakers and even the exportation of scrap glass (cullet) were prohibited under penalty of death [4]. Other countries, particularly England and France, actively sought the Venetian glassmaking knowledge and artisans were offered both bribes and promises of protection if they would leave Venice to set up rival glassworks elsewhere. Besides limiting travel, the Senate also used rewards in efforts to keep critical knowledge safe in Venice and in 1376 declared that all Murano glassworkers be declared burgher of Venice (a member of its middle class). This meant that noblemen could now marry the daughters of master glassmakers and the children of such marriages would be noble [21, 23]. No other tradesman's daughters were eligible for such a high honor. Nevertheless, while the noble character of glass-working was recognized, giving glassworkers elevated status, they were still virtually prisoners and not allowed to leave the island of Murano [21].

Murano glass production continued to grow during the fifteenth century and Venice emerged as one of the five most powerful states in Italy [1]. During this time glassworking techniques improved further, probably heavily influenced by events in the eastern Mediterranean. In particular, it is believed that the destruction of the city of Damascus (a great glass center) by the Mongolian warlord Tīmūr the Great in 1400 led to an influx of craftsmen to Italy [22, 23]. A similar influx may also have happened at the end of the Byzantine Empire when Constantinople fell to the Turks in 1453 [22]. By the year 1490, the glassmaking industry had become of such financial importance to the Venetian Republic that it was placed under the authority of its special Council of Ten [21], which was tasked with maintaining the security of the Republic and preserving the government from overthrow or corruption.

During this same time period, the Venetians succeeded in making what is often considered their highest quality glass, referred to as *cristallo*. Cristallo was first mentioned in 1409 [22, 27] and was well known by the mid-fifteenth century [1]. Cristallo was an usually clear, colorless, and hard glass that was free from flaws while providing an extremely long working time, although this "colorless" glass was neither as transparent nor as brilliant as our common modern glass [1, 21, 29]. The working time is the time between its removal from the heat of the furnace and the point when it becomes too rigid to continue shaping without re-heating it [1]. Cristallo was strong, extremely malleable and could be worked very thin, making objects seem almost weightless. This combination made it a unique glass for the fifteenth century and enabled glassworkers to make wonderfully elegant and intricate forms [1, 22, 27].

4.4 Materials and Methods

As stated above, one of the factors that contributed to the dominance of the Venetian glass industry was the quality of raw materials available to the glass-makers of Murano. Such materials were easily obtainable and included critical components such as soda ash from Egypt or from Alicante in Spain, as well as silica from Lido (a sandbar near Venice) or Verona. Other raw materials utilized by the glass industry included wood from Eastern Venetia and the lower Alps, clay from Vicenza, and salt from Dalmatia [26].

Until the beginning of the fourteenth century, the nearly exclusive source of silica used by the Venetian glassmakers was various local sands referred to as *tera* or *terra* [30, 31]. Sources of these sands included adriatic coastal sites (Pola, Friuli) as well as other nearby Italian sites (Polcenigo, Vicenza and Trapani) [31]. In addition to silica, these sands are thought to have provided considerable alumina, as well as iron oxide, lime, magnesia, and frequently small amounts of manganese [8, 25].

It had long been known that the cleaner and whiter the silica used to form the basis of the glass batch, the clearer the resulting glass. As a result, these sands were gradually replaced with *cogoli*, a term used to refer to flint pebbles (a form of the mineral quartz) obtained from the beds of either the river Ticino and the river Adige near Vicenza and Verona [1, 29–33]. Before use, these pebbles were calcined (heated red-hot in an oxidizing atmosphere) and then thrown into pure water so that the rapid temperature change would assist with cracking and breaking the pebbles [31, 32]. The resulting material could then be ground and sieved to form a fine quartz powder that was purer than the previously used sands [1, 8, 26, 31, 32]. The resulting material was $\sim 98\%$ silica, which reduced impurities that contributed to coloring of the glass (iron, chromium, etc.) [29–31], and provided $\sim 70\%$ of the typical glass batch [1]. The pebbles from Ticino were considered to be superior to those of Verona, which produced a yellow tinted glass [32]. These flint pebbles were used in Murano as early as the fourteenth century and became the near exclusive silica source of the Venetian glassmakers for the next several centuries [30, 32].

While careful selection of silica source could affect the resulting glass composition, it could be argued that the largest difference between the glass compositions of the previous Roman period and the improved Venetian material was due to the alkali source. As discussed in Chap. 3, most Roman glasses were prepared with *natron* as the alkali of choice [8, 29]. The Venetians, however, favored the use of plant ashes, in particular the ashes imported from the Levant (modern Syria, Israel, Lebanon, and the Sinai in Egypt) [11, 29–32, 34, 35]. It has been proposed that Venice's close economic ties to the Levant, particularly the Syrian glassmaking knowledge and technology shared with Venice in the thirteenth century, resulted in this blending of the plant ash-based methods of the Levant with previous Roman glassmaking techniques [1, 24, 29]. However, knowledge gained from Byzantine and Islamic glassmakers could also have significantly contributed to this choice in alkali.

The soda ash imported from the Levant originated from the burning of certain plants that grew in saline soils, such as the fringes of the desert or along the sea coast [35]. These plants are thought to have belonged to the large family of the *Chenopodiaceae*, in particular the plant *salsola kali* (Fig. 4.3) [11, 29–32, 34, 35].

Fig. 4.3 *Salsola kali* [Botanical print (dated 1792); photo by Júlio Reis (2004)]

In the literary Arabic, these plants were called *ushnān*, but were called *keli* or *kali*[3] in the vernacular [31, 35], this latter term being used to refer to both the plant and its ash [36]. The majority of this ash came from plants that were collected by the Bedouin in the Syrian Desert. These plants would then be burnt to produce ash in the form of solid lumps or pebbles and sold through the Levantine markets [35].

These Levantine ashes, referred to in Venice and Murano as *allume catino*,[4] were in common use by 1285 [30] and were used almost exclusively in Murano until the end of the 1600s [34]. During this time their use was even protected by the Venetian government, and the use of other plant ashes for glassmaking was expressly prohibited [28, 35]. *Allume catino* had relatively high soda content (as high as 30–40%), as well as quite large amounts of potassium, calcium, and magnesium carbonates [11, 30, 32, 34, 35]. However, as shown in Table 4.2, the exact source of the ash could play a factor in its chemical composition. For example, the ash of Syria was regarded to be better than that of Egypt, as the Syrian ash was blacker in color because of its higher carbon content [32, 34, 35]. Higher carbon content meant that the ash consisted of greater amounts of carbonate and bicarbonates which would dissociate to form oxides and thus be more readily incorporated into the glass [20].

In addition to *allume catino*, the Venetian glassmakers also utilized *barilla* or Spanish ash, obtained from the burning of marine plants (*salsola sativa, halogeton*

---

[3] Addition of the Arabic article *al-* gave *al-qalīy*, the Arabic term for the ash. This eventually became the familiar *alkali*, the modern term used to refer to the bases and elements isolated from these ashes. The Latin form of kali, *kalium*, is also the Latin name of potassium and the source of the symbol K for this element.

[4] In addition to the term *allume catino*, the Leventine ashes were referred to by the alternate terms *allume catina, lume catino, lume gatino, cenere di Levante* (ash of Levent), *rocchetta* (in pieces), and *polverino* (in powder form) [32, 34, 35].

Table 4.2 Composition of *salsola kali* plant ash from various sources

Component[a]	Levant (Israel)[b]	Sicily (Italy)[c]	Attica (Greece)[d]	Crete (Greece)[d]	Pembrokeshire (UK)[d]
Na_2O	14.3	17.0	23.1	19.1	15.2
K_2O	15.5	9.0	19.0	22.2	33.4
CaO	14.4	16.0	14.4	15.5	13.8
MgO	9.0	11.0	6.6	6.6	6.2
P_2O_5		2.4	1.0	1.3	1.5
SO_3	5.5	2.2	2.1	1.3	2.6
SiO_2	<1.0	6.8	7.0	2.0	5.0
Al_2O_3		1.2	1.1	0.5	0.5
Fe_2O_3		1.2	0.4	0.1	0.2
% Carbonate	77.6	91.4	33.8	15.7	44.2

[a] % by weight. % carbonate refers to the percentage of total alkali moles present as carbonate
[b] Ref. [35]
[c] Ref. [28]
[d] Ref. [20]

sativus, salsola kali, and *suaeda maritima*) from the salt marshes of Alicante, Spain and other parts of the Mediterranean [11, 32, 37]. The highest quality *barilla* for glassmaking was called *agua azul,* of which Alicante was the sole source [37]. The source of this form of *barilla* was described as a shrub with blue green berries, thought to be *salsola sativa* [32], which gave this particular *barilla* ash a blue color [37]. Like *allume catino, barilla* was a soda-rich ash (up to 25–30%) containing significant quantities of calcium salts [11, 31, 35]. While it is believed that both the Levantine and some *barilla* ashes may have been derived from *salsola kali,* it is important to remember that the chemical composition of the plant ash is largely dependent on the soil in which the plants grow, as illustrated by the data given in Table 4.2. Thus, while the Levantine and *barilla* ashes may have been similar, they were distinct raw materials with differing characteristics. Of the two, *allume catino* was the initial and preferred material with *barilla* becoming more common by the sixteenth century [11, 37]. Neri later wrote of both materials, also expressing a preference for the Levantine ash over *barilla* [38]:

> Barillia of Spain, though it be usually fatter than the former, yet it makes not a glass so white and fair as that of the Levant, because it always inclines a little to an azure colour.

As stated above, this preference was due to the fact that glass from *barilla* would tend to suffer from some light blue coloring [32, 37, 38], which has been proposed to be a result of iron oxide content in the ash [32].

The Venetians utilized lead as a stabilizer and it was introduced to the glass batch in several forms. Most common was red lead, referred to as *minio* (Pb_3O_4), with alternates including litharge (PbO) and white lead (basic lead carbonate, $Pb_3(CO_3)_2(OH)_2$) [31, 32]. Although calcium sources such as limestone ($CaCO_3$) and dolomite ($CaMg(CO_3)_2$) were available to the Venetian glassmakers, they do not seem to have been used as stabilizers [31] and all calcium content in Venetian glass was introduced as impurities contained in the silica or alkali sources [30]. This is consistent with the belief that the importance of lime was not yet recognized at this point in time and thus calcium was not intentionally added as a major constituent before the end of the seventeenth century [8, 11, 30, 39].

Another significant contribution to the advancement of the Venetian methods was the introduction of new processes for the preparation of the alkali raw materials [35]. The plant ash (either *barilla* or *allume catino*) was shipped to Venice as hard pieces of calcined residue. Although these chunks of calcined residue then required pulverization after arrival, it was preferred over ash that arrived in powdered form [25, 30, 34]. The pulverized ash was then purified by a series of sieving, filtering, and/or recrystallization steps, which could remove unwanted impurities such as iron and aluminum-containing species [8, 11, 29]. The methods were being practiced in Italy by the fifteenth century [35]. Neri later gave considerable attention to alkali preparation, discussing in great detail the purification of plant ashes to prepare good alkali salts for clear crystal glass. This process consisted of repeated sieving the raw ash, dissolving it in boiling water combined with *Tartar of red wine* (potassium bitartrate, $KHC_4H_4O_6$), filtering the mixture, and evaporation of the water to obtain the purified material [8, 38].

It was such purification of the plant ash that has been considered to be the main secret in obtaining the clear, colorless nature of cristallo [30] and it is thought that the combination of silica derived from the flint pebbles of the river Ticino and such highly purified ash allowed its successful production [1, 29, 32]. The combination of these two highly pure components would contribute few color-causing impurities, thus needing very little decolorant (MnO_2) to produce a colorless glass [30]. This is supported by the fact that analysis of cristallo glass has shown Fe_2O_3, Al_2O_3, and MnO content below 1.0, 0.3, and 0.5% respectively [29].

This purification, however, had to result in some difficulties for the glassmaker in terms of stability of the glass. The process of purification of the ash greatly reduced the alkaline-earth (calcium and magnesium) and alumina content, due to the low solubility of these salts [8, 30]. As a consequence, the glass obtained from the purified ash contained low amounts of stabilizers and was of significantly reduced chemical stability [8, 30]. This is definitely observed in many examples of Venetian glass of that period now in museums, which have developed surface crizzling (a multitude of fine surface fractures) because of their poor chemical durability. Some such extreme specimens are even sticky to the touch and appear to sweat [8]. Later recipes for cristallo, however, seem to describe a lead glass, which may contradict the belief that cristallo was a simple soda-lime glass. It has been suggested by some that perhaps the view of cristallo as a soda-lime glass applies to that used in artistic blown glass, but that used in most other cases was a lead-based glass [31].

References

1. Cummings K (2002) A history of glassforming. A & C Black, London, pp 108–120
2. Asimov I (1991) Asimov's chronology of the world, the history of the world from the big bang to modern times. HarperCollins, New York, pp 100–101
3. Haywood J (2002) The atlas of past times. The Brown Reference Group, London, pp 71–79
4. Philips CJ (1941) Glass: the miracle maker. Pitman Publishing Corporation, New York, pp 3–14
5. Brill RH (1963) Ancient glass. Sci Am 109(5):120–130
6. Lambert JB (1997) Traces of the past. Unraveling the secrets of archaeology through chemistry. Addison Wesley, Reading, pp 124–127
7. Wedepohl KH, Simon K (2010) The chemical composition of medieval wood ash glass from Central Europe. Chemie der Erde 70:89–97
8. Kurkjian CR, Prindle WR (1998) Perspectives on the history of glass composition. J Am Ceram Soc 81:795–813
9. Lambert JB (2005) The 2004 Edelstein award address, the deep history of chemistry. Bull Hist Chem 30:1–9
10. Charleston RJ, Angus-Butterworth LM (1957) In: Singer CJ, Holmyard EJ, Hall AR (eds) A history of technology, vol 3. Oxford University Press, New York, pp 206–209
11. Turner WES (1956) Studies in ancient glasses and glassmaking processes. Part V. Raw materials and melting processes. J Soc Glass Technol 40:277T–300T
12. Sayre EV, Smith RW (1961) Compositional categories of ancient glass. Science 133:1824–1826
13. Derry TK, Williams TI (1961) A short history of technology, from the earliest times to A.D. 1900. Oxford University Press, New York, p 85
14. Morey GW, Kracek FC, Bowen NL (1930) The ternary system K_2O-CaO-SiO_2. J Soc Glass Technol 14:149–187
15. Tait H (ed) (1991) Glass, 5,000 years. Harry N. Abrams, Inc., New York, p 179
16. Asimov I (1991) Asimov's chronology of the world, the history of the world from the big bang to modern times. HarperCollins, New York, pp 119–122
17. Partington JR (1999) A history of Greek fire and gunpowder. Johns Hopkins, Baltimore, pp. 12–18, 28–32
18. Shortland AJ, Tite MS (2000) Archaeometry 42:141–151
19. Lambert JB (1997) Traces of the past. Unraveling the secrets of archaeology through chemistry. Addison-Wesley, Reading, pp 109–116
20. Tite MS, Shortland A, Maniatis Y, Kavoussanki D, Harris SA (2006) The composition of the soda-rich and mixed alkali plant ashes used in the production of glass. J Archaeol Sci 33:1284–1292
21. Roger F, Beard A (1948) 5,000 years of glass. J. B. Lippincott, New York, pp 28–38
22. Macfarlane A, Martin G (2002) Glass: a world history. University of Chicago Press, Chicago, pp 10–26
23. Sarton G (1947) Introduction to the history of science. The William & Wilkins Co., Baltimore, vol III, part I, pp 170–173
24. Jacoby D (1993) Raw materials for the glass industries of Venice and the Terraferma, about 1370–about 1460. J Glass Studies 35:65–90
25. Haywood J (2000) Historical atlas of the medieval world, AD 600–1492. Barnes & Noble, New York, p 3.12, 3.17
26. Sarton G (1931) Introduction to the history of science. The William & Wilkins Co., Baltimore, vol II, part II, pp 1040–1041
27. Axinte E (2011) Glass as engineering materials: a review. Mater Des 32:1717–1732
28. Gies F, Gies J (1994) Cathedral, forge, and waterwheel. Technology and invention in the Middle Ages. HarperCollins Publishers, New York, p 102
29. Mass JL, Hunt JA (2002) The early history of glassmaking in the Venetian lagoon: a microchemical investigation. Mater Res Soc Symp Proc 712:303–313

30. Verità M (1985) L'invenzione del cristallo muranese: una verifica analitica delle fonti storiche. Rivista della Staz Sper del Vetro 15:17–29
31. Moretti C (1985) Glass of the past. Chemtech 15(6):340–344
32. Moretti C (2001) Le materie prime dei vetrai veneziani rilevate nei ricettari dal XIV alla prima metà del XX secolo. II Parte: elenco materie prime, materie sussidiarie e semilavorati. Rivista della Staz Sper del Vetro 31:17–32
33. Neri A, Merrett C (2003) The world's most famous book on glassmaking, the art of glass. The Society of Glass Technology, Sheffield, p 65
34. Zecchin P (1997) I fondenti dei vetrai muranesi. I Parte: l'allume catino. Rivista della Staz Sper del Vetro 27:41–54
35. Ashtor E, Cevidalli G (1983) J Euro Econ Hist 12:475–522
36. Ruland M (1984) A lexicon of alchemy. (trans: Waite AE) Samuel Weiser, York Beach, p 186
37. Zecchin P (1997) I fondenti dei vetrai muranesi. II Parte: gli scritti dei secoli XV, XVI, XVII. Rivista della Staz Sper del Vetro 27:251–265
38. Neri A, Merrett C (2003) The world's most famous book on glassmaking, the art of glass. The Society of Glass Technology, Sheffield, pp 59–63, 75, 98
39. Turner WES (1956) Studies in ancient glasses and glassmaking processes. Part III. The chronology of the glassmaking constituents. J Soc Glass Technol 40:39T–52T

Chapter 5
Applications to Chemical Apparatus

As discussed in Chap. 1, glass has been used extensively in modern science, particularly in terms of laboratory glassware. In fact, the image of glass objects such as distillation heads, beakers, flasks, vials, and test-tubes has become mainstay in the public's common view of chemical laboratories and the practice of chemical research. Even for the practitioner of science, it is hard to imagine what the chemical laboratory would be like without glass. After all, no other material can really match its combination of low cost, chemical stability, thermal durability, as well as freedom and versatility in the design of chemical apparatus for nearly any desired application. The importance of glass to the chemical arts was even highlighted by Neri in his *Art of Glass*, stating [1]:

> It is a thing profitable in the service of the Art of distilling, and Spagyrical, not to say necessary to prepare Medicines for man, which would be impossible to be made without the means of Glass, so that herewith are made so many sorts of Instruments, and Vessels, as Bodies, Heads, Receivers, Pelicans, Lutes, Retorts, Athenors, Serpentines, Vials, Cruces, square and round Vessels, Philosophical Eggs, Globes, and infinite other sorts of Vessels, which every day are invented to compose and make Elixars, Arcana, Quintessences, Salts, Sulphurs, Vitriols, Mercuries, Tinctures, separation of Elements, all Metalline things, and many others…

Chemical apparatus such as that described by Neri date back to the end of the first century [2–4], at a time in which glass was rising to its initial heights in the Roman Empire. Thus, it is somewhat surprising that glass-based apparatus for the chemical arts do not seem to have been developed to any significant degree during the Roman period [5, 6]. The late use of glass for chemical equipment stems from the fact that while typical soda-lime glasses can provide low melting materials that are easy to work with, they usually lack sufficient chemical durability to be practical [7]. In addition, laboratory glassware must often withstand severe temperature changes in the presence of strong reagents. Hence, for laboratory glassware to be useful, it must not only be resistant to chemical attack, but must also be durable under thermal stress [6]. The combination of poor quality and the thick, irregular nature of the many

S. C. Rasmussen, *How Glass Changed the World*,
SpringerBriefs in History of Chemistry, DOI: 10.1007/978-3-642-28183-9_5,
© The Author(s) 2012

medieval glasses resulted in the frequent breaking of the vessels during chemical applications such as distillation [8].

5.1 Chemical Durability

The low chemical durability of typical soda-lime glasses is largely due to the high sodium oxide content [7, 9]. For example, glass made entirely of silica and soda is actually quite soluble in water. The action of water on glass, however, is not a simple solute–solvent interaction. The process is somewhat complicated and involves first adsorption and then absorption of the water by the glass. The absorbed water then results in hydrolysis that solublizes glass, as well as intro-duces new chemical species that can further degrade the material. The process is more nearly akin to diffusion and disintegration than it is to true solubility [10–12].

A more detailed, yet qualitative, description of the action of water on a typical soda-lime-silica glass is as follows [11, 12]. Water is first absorbed, which can result in ion exchange. As shown in Scheme 5.1, this process involves the removal of alkali and alkaline earth ions which are replaced by protons or hydronium ions [12].

In addition to the exchange and removal of alkali ions, it is also possible to leach smaller sodium silicate species from the glass. The sodium silicates leached from the glass then undergo hydrolysis to produce sodium hydroxide and colloidal silicic acids (Scheme 5.2). In both of these cases, a leached layer is formed, often referred to as hydrogen glass. The concentration of the network modifiers such as sodium or calcium is greatly depleted in this external layer, increasing the relative concentration of silica [12].

Depending upon the conditions involved, the solution of sodium hydroxide resulting from these processes may or may not be concentrated enough to result in significantly high pH solutions. At pH values above ten, the hydroxide ions can start to break Si–O bonds in the glass matrix, causing the breakdown of the silica backbone (Scheme 5.3) [12]. Such attack by hydroxide on the glass surface can cause a permanent dull stain [11].

If CO_2 is also present, further reactions can occur between the gas and sodium species, resulting in the formation of sodium carbonate (Scheme 5.4). Needle-like crystals of the carbonate salt can sometimes be found as a white deposit on the glass surface. In some cases, this can be seen as a delicate tracery of branching crystals. Washing the glass surface can remove such salt deposits, but does not remove the surface layer of silica obtained by the initial leaching and hydrolysis, thus permanently dulling the brilliancy of the glass [11].

This overall disintegration process will tend to continue until the glass is completely decomposed, although the rate of decomposition will depend upon a number of factors. One such critical factor is the composition of the glass. The other components of the glass, silica and lime, are relatively insoluble and thus can provide a stabilizing effect against the attack of chemical reagents on the sodium oxide [11, 12]. As a consequence, the chemical resistance of glass can be increased

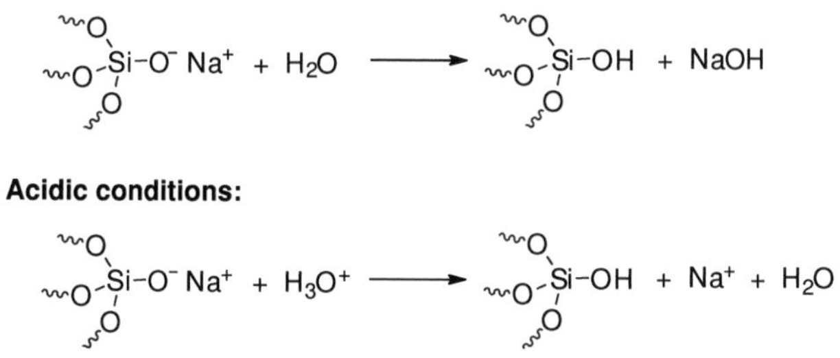

Neutral conditions:

Acidic conditions:

Scheme 5.1 Ion exchange of alkali ions

Scheme 5.2 Hydrolysis of sodium silicates

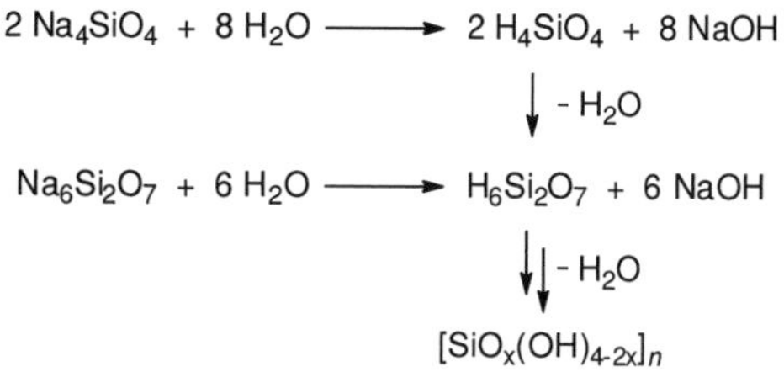

$$2\,Na_4SiO_4 + 8\,H_2O \longrightarrow 2\,H_4SiO_4 + 8\,NaOH$$
$$\downarrow -H_2O$$
$$Na_6Si_2O_7 + 6\,H_2O \longrightarrow H_6Si_2O_7 + 6\,NaOH$$
$$\downarrow\downarrow -H_2O$$
$$[SiO_x(OH)_{4-2x}]_n$$

by decreasing the soda content and increasing lime. Unfortunately, this is not a viable solution as this also increases the tendency toward devitrification (i.e. glass crystallization resulting in frosting and loss of transparency) [7]. It has also been shown that the chemical durability can often be increased by adding alumina, substituting lead oxide (PbO) for lime, substituting zinc oxide (ZnO) for alkali, or by using a combination of soda and potash rather than either alkali base alone [11]. It is also now understood that the problem of devitrification caused by high amounts of less soluble oxides can by overcome through the application of multiple oxide components. For example, the addition of magnesium oxide to typical soda-lime compositions can result in a more chemically durable glass without devitrification [7].

   In 1926, as an attempt to better understand the effect of the stabilizing oxides utilized in glassmaking, glass historian W. E. S. Turner[1] prepared a series of tertiary glasses of the general formula $6SiO_2 \cdot (2-n)Na_2O \cdot nMO_x$ [9, 10]. Eight different oxides (MO_x) were utilized and with each set of glasses, the content of

---

[1] William E. S. Turner (1881–1963). In 1915, he was invited to become the Head of the new Department of Glass Technology at the University of Sheffield, a position which he held until 1945. Recognizing the importance of drawing together the various groups of people interested in glass into one organization, he founded The Society of Glass Technology in 1916. He served as president of the International Glass Society from 1933 to 1953. He never lost his interest in the historical aspects of glass and continued to publish as late as 1962 [13, 14].

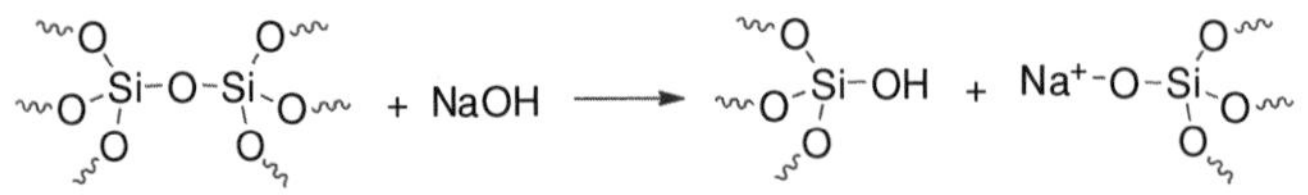

Scheme 5.3 Cleavage of the silica backbone

Scheme 5.4 Production of sodium carbonate

$$Na_2O + CO_2 \longrightarrow Na_2CO_3$$

$$NaOH + CO_2 \longrightarrow NaHCO_3 \xrightarrow{NaOH} Na_2CO_3 + H_2O$$

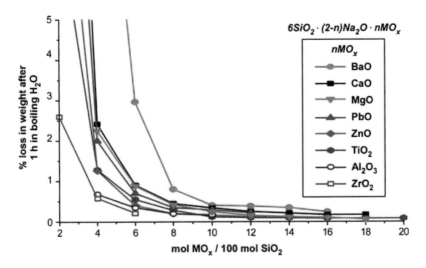

Fig. 5.1 Water resistance of tertiary oxide glasses as a function of MO_x content (data from [9])

the stabilizing oxide was varied to give a range of glasses of different chemical composition. The complete series of glasses was then treated with boiling water to evaluate their water resistance. The results of these tests are shown in Fig. 5.1. In all cases, the replacement of Na_2O by even a small amount of one of the other oxides results in a rapid increase in chemical stability [9, 10]. In fact, lime or magnesium oxide content as low as 3% results in significant increases in chemical resistance [14]. However, some oxides introduce greater stability than others, with a rough trend that the application of oxides with a very low solubility in water result in more water resistant glasses [10].

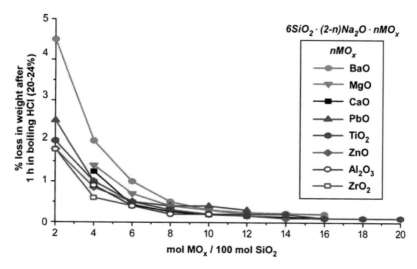

Fig. 5.2 Acid resistance of tertiary oxide glasses as a function of MO_x content (data from [9])

As pointed out above, the water resistance of glass can also be enhanced by using mixtures of soda and potash, rather than either alkali alone. The extent of resistance, however, is dependent on the exact ratio of alkali oxides. Thus, the resistance to the action of water is maximized in glasses with a sodium oxide to potassium oxide ratio of 3:7 [10].

The action of neutral salt solutions on glass is essentially the same as that of water itself. This is also true for dilute acid solutions such as nitric and sulfuric, and much evidence suggests that their effect may be wholly due to the water present [10, 11]. In fact, concentrated sulfuric acid has been found in some cases to have a smaller corrosive action than pure water [10]. As can be seen in Fig. 5.2, all of the glasses exhibited greater resistance to HCl than pure water, with overall stability again increasing as greater amounts of Na_2O are replaced with less soluble oxides [9, 10].

Basic solutions usually result in more significant disintegration than that caused by water or dilute acids [10, 11]. Thus, as shown in Fig. 5.3, much larger amounts of material are lost by the action of either Na_2CO_3 or NaOH solutions [9, 10]. This is not completely unexpected as most glasses consist of $\sim 70\%$ silica or other acidic oxides [10]. Less expected is perhaps the fact that while only mildly alkaline in comparison to NaOH, the degree of corrosion produced by Na_2CO_3 solutions is greater on some glasses that the action of hydroxide solutions [9]. This can be seen in Fig. 5.3 in which glasses with low amounts of MO_x exhibit greater corrosion in the carbonate solution in comparison to the same glasses in comparable hydroxide solutions. This is somewhat inverted for glasses with higher MO_x concentrations due to the fact that additional MO_x content provides good resistance to the carbonate solution while providing only moderate protection to hydroxide. Overall,

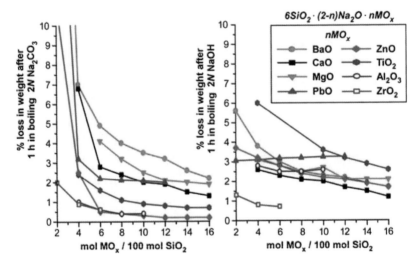

Fig. 5.3 Base resistance of tertiary oxide glasses as a function of MO_x content (data from [9])

the effect of MO_x on the base resistance of the resulting glass is generally related to oxide basicity, with more basic oxides giving higher resistance, while more acidic oxides are more readily attacked by caustic solutions [9]. Of the two most common stabilizing oxides found in the glasses discussed in the previous chapters, it has been shown that either lime (CaO) or magnesia (MgO) results in increased chemical resistance, although neither are as effective as many of the other oxides analyzed in Figs. 5.1, 5.2 and 5.3 [9, 10, 15]. Between lime and magnesia, magnesia is more beneficial in small concentrations for increasing water resistance. The difference between the two decreases at higher concentrations, but magnesia still gives overall superior resistance in comparison to lime. In contrast, however, the difference between the two oxides in relation to increasing acid resistance is very small [15].

Although magnesium appears to be the most beneficial oxide for improved water and acid resistance, lime imparts a markedly improved resistance to alkaline solutions in comparison to magnesium. In fact, the mass loss from magnesia glasses due to caustic solutions is two and half to three times that of the lime glasses [15]. As a consequence, glasses containing mixtures of the two oxides would perhaps give the best overall compromise in terms of broad chemical stability.

5.2 Thermal Durability

The second critical property for laboratory glassware is its thermal durability, of which the primary determining factor is its thermal expansion [10, 16]. Like most solid materials, glass undergoes thermal expansion that can result in increased

Table 5.1 Comparative expansion factors for various oxides

Oxide[a]	Constant	Oxide[a]	Constant
Na_2O	1296	ZrO_2	69
K_2O	1170	Al_2O_3	52
BaO	520	ZnO	21
CaO	489	SiO_2	15
MgO	135		

[a] Values factored by 10^9 for clarity. Ref. [16]

stress during rapid temperature changes and the ability to withstand this thermal shock is important in many applications of glass [16, 17]. In particular, chemical glassware and glass bakeware both depend largely upon this ability [17].

Determining the net thermal durability of a particular glass is complicated as a number of factors can play a role, including the glass tensile strength, manufacturing methods and handling, glass thickness, and sites of stress. However, the general rule is that the lower the expansion coefficient of a glass, the greater its thermal endurance [17]. For example, a glass with an expansion coefficient of 3.2 ppm/°C can withstand three times the temperature change required to break a glass with an expansion coefficient of 9.6 ppm/°C.

For the majority of glasses studied, the relationship between thermal expansion and chemical composition of the glass appears to be relatively simple and approximately linear [16]. The cubic (or volumetric) thermal expansion coefficient (3α) of a particular glass can be calculated using the following relationship

$$3\alpha = \sum k_n p_n = k_1 p_1 + k_2 p_2 + k_3 p_3 \ldots$$

where p_n represents the percentage by weight of each constituent oxide in the glass and k_n are constants that relate to the contribution to expansion by that oxide. These constants are referred to as expansion factors and represent the contribution that each 1% of the oxide makes to the total cubic expansion of the glass [10, 16]. Comparative constants for a number of oxides are given in Table 5.1.

Due to the very high expansion factor for sodium oxide, the thermal expansion of simple soda glasses increases with soda content, thus resulting in glasses with low thermal durability and a tendency to break under rapid heating. As shown in Fig. 5.4, the substitution of sodium oxide by another oxide results in decreased thermal expansion with increasing content of the new oxide. Both soda and potash exert such a dominating effect on the expansion of the glass that slight variations in chemical composition, in some cases of even less than 1%, can cause a significant variation in the thermal expansion [16].

As can be seen in Fig. 5.4, the majority of oxides studied have very similar effects on reduction of thermal expansion. The exceptions include barium oxide (BaO) and lead oxide (PbO), both of which provide somewhat small reductions in glass expansion and are thus poor oxides for terms of increasing thermal durability. The other important exception is magnesia, which provides by far the best reduction in thermal expansion. Thus, for the two most common stabilizing oxides found in the

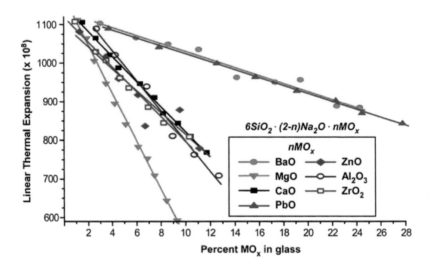

Fig. 5.4 Linear thermal expansion of tertiary oxide glasses as a function of MO_x content (data from [16])

glasses discussed in the previous chapters, both lime and magnesia would provide significantly decreased thermal expansion. However, it would be additional magnesium oxide content that would have the most significant effect [16].

Other important factors that contribute to overall thermal durability include the thickness. This is due to the fact that thicker glass provides more glass to expand and thus greater overall cubic expansion. As such, the thicker the glass, the more likely it is to break as a result of the stress induced by expansion. Lastly, it is also understood that glasses can withstand much greater temperature differences when suddenly heated than when suddenly cooled [17].

5.3 Increased Durability of Venetian Glass

The preparation and purification methods employed by the Venetian glassmakers ensured raw materials with a more consistent composition, resulting in the production of a more consistent and uniform glass. However, the use of the Levantine and *barilla* ashes over *natron* may have played the greatest role in the improvements in glass, as these alkali sources contributed increased amounts of lime, magnesia and potash.

As shown in Table 5.2, the analysis of Venetian glass samples from the sixth to fifteenth centuries reveals decreasing soda content after the eleventh century, which is coupled with increasing calcium content [18–20]. In addition, the later glasses also exhibit significant increases in both magnesium and potash. While the dating of these samples is somewhat broad, the time span does encompass the

Table 5.2 Chemical composition of Venetian glass samples from the sixth to fifteenth century

Content[a]	Sixth to eighth century[b]		Eleventh to fourteenth century[c]			Late fifteenth century[b]		
	A	B	C	D	E	F	G	H
SiO_2	69.1	69.4	68.5	68.6	70.0	64.9	65.5	64.9
Al_2O_3	2.50	3.06	1.95	1.40	1.90	1.31	1.73	1.68
Na_2O	16.0	17.6	12.5	12.5	11.7	12.7	10.8	11.6
K_2O	0.58	0.74	3.00	2.90	1.45	2.72	1.96	2.45
CaO	7.8	6.7	8.20	9.05	11.9	11.26	12.15	11.86
MgO	0.60	0.66	2.70	3.05	1.15	4.42	4.54	4.36
Fe_2O_3	0.40	0.41	0.47	0.38	0.30	0.604	0.717	0.712

[a] % by weight
[b] Ref. [20]
[c] Ref. [19]

period believed to correlate to the marked improvement in glass technology (i.e. the later thirteenth century) [6, 21–23], and the changes in the composition of these later samples are consistent with the discussed improvements. In addition, it is reasonable to argue that the higher content of these elements is due to the use of the Levantine or *barilla* ashes that are known to contain significant amounts of calcium, magnesium, and potash [19, 23–27].

The preparation and purification methods utilized would have removed insoluble, non-fusible components from the resulting glass products, which otherwise would have acted as points of stress during rapid heating. In addition, the reduced soda content combined with the higher content of the stabilizing oxides would result in a material that exhibited both higher chemical durability and less thermal expansion [7, 9, 15, 16]. As luck would have it, glasses of a chemical composition that provides the greatest resistance to the action of water and acids are also typically those which are the best heat-resistant glasses [10].

As a result, the improved Venetian glass would therefore be more resistant to the action of water, acids, and bases, and would be less affected by rapid temperature changes, thus making it ideal for use in laboratory glassware. The introduction of this improved glass then paved the way for new and varied applications of its use. In particular, it has been argued that the development of both laboratory glassware and lenses were a result of these Venetian advancements [6, 28].

5.4 Chemical Glassware

Most early chemical vessels and apparatus most likely developed from kitchenware. As such, it is not too surprising that the terms used for such early vessels were primarily derived from the words for jars, pots, cups, urns, and kettles [29]. By the second century, apparatus such as beakers, jars, flasks, mortars and pestles, funnels, sieves, filters, and crucibles were all in use by practitioners of the chemical arts [4, 30]. Among the many specialized forms of vessels used for chemical storage and operations was the *phial*, often mentioned as a long-necked flask, and the *urinal*, a flask which had a wider neck. The latter was frequently used in medieval medicine for the analysis of urine [8]. The phial could have been the precursor to the later, similarly designed Florence flask, which was commonly used in the seventeenth century [31].

Early laboratory glassware was typically fabricated from basic soda-lime glasses which were relatively high in silica and low in alkali, as such compositions were more resistant to chemical attack. However, such glasses were difficult to work and still had somewhat high expansion coefficients. As a result, articles of glass had to be made very thin in order to increase their thermal durability, which made glassware of this type correspondingly fragile [32].

In contrast, the chemical composition of the newer Venetian glasses provided materials that were both chemically and thermally durable, which made it especially practical for laboratory apparatus. As a consequence, the flourishing industry at Venice and Murano greatly influenced chemical progress [33, 34]. The ability to produce more laboratory apparatus and vessels from glass allowed much greater freedom and versatility in their design [28]. No matter what shape was needed, it could be made of glass, and it is felt by some that the scientific preeminence of Italy between 1550 and 1650 is directly related to the ascendency of Venetian glass [32]. At the same time, the Venetian glass industry also received great impetus from the growing general use of glass for chemical vessels.

5.5 Stills

The ability to use glass in the production of laboratory apparatus allowed much greater freedom and versatility in design, and nowhere was this more evident than in the rapid evolution of the still. The still, thought to have been the earliest specifically chemical instrument, dates back to the end of the first century (Fig. 5.5) [2–4]. It is Maria the Jewess[2] who first described this distilling apparatus with its development already fairly advanced in her writings. As such, she is generally given credit for its invention [2, 4, 29, 35, 36].

As shown in Fig. 5.6, the form of the traditional still consisted of three components: the distillation vessel (*cucurbit*), the still head (*ambix*) with an attached delivery tube (*solen*), and the receiving vessel (*bikos*). These receiving vessels were generally comprised of a small body with a long thin neck and, as discussed above, such flasks in general were also called a *phial* [29]. The term *ambix* was later transformed through the addition of the Arabic article *al-* to become *alanbîq*,

---

[2] Also known sometimes as Mary or Miriam, Maria the Jewess was an alchemist of the first century CE. Little is known about her other than writings ascribed to her, which survives only in quotations by the later alchemist Zosimos. She is also credited with the invention of the water-bath and the *kerotakis* apparatus. She is alleged by some to be the sister of Moses [2, 35, 36].

Fig. 5.5 Early woodcut of a
typical still

Fig. 5.6 Basic components
of the early still (Ref. [6],
Reproduced with permission
of the Division of the History
of Chemistry of the American
Chemical Society)

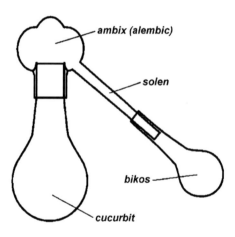

which eventually became *alembic*.[3] By the Middle Ages, the term *alembic* was
used to refer to the still as a whole [2–4, 8].

The early stills were made from a mixture of primarily earthenware (with the
interior glazed) and copper, although sometimes glass receiving vessels were used
[2, 3, 8]. As glass industries evolved, it became more common to use glass for first
the *alembic* and then later for both the *cucurbit* and *alembic* [8]. One of the

---

[3] Alternate forms included *alembik, alembyk, alembike, alembyke, alimbeck, alembeke,
alimbecke, alimbeck,* and *limbick* [8].

Fig. 5.7 External cooling trough as depicted in the 1420 medical treatise of John Wenod

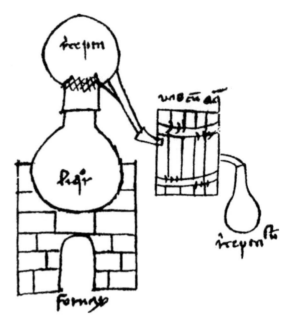

difficulties encountered with the use of glass in still components was the breaking of the vessels because the glass was typically thick, irregular, and of poor quality. In fact, Pliny the Elder stated that [37]:

> Glass is unable to stand heat unless a cold liquid is first poured in.

To counter this, a thick coating (up to two or three fingers) of clay was applied to the exterior of the *cucurbit* [8, 38]. This helped reduce breaking, but the poor heat transmission of the coating resulted in unnecessarily long preheating periods. This combination of poor heat transmission and the lack of efficient cooling made it difficult to distill volatile liquids such as alcohol [39].

Initial efforts to improve cooling methods were to cool the delivery tube (*solen*) with wet sponges or rags. As the delivery tube was now typically cooler than the still-head, condensation occurred primarily in the delivery tube. Because of this, the typical medieval *alembic* no longer contained an inner rim for collecting the condensate within the still head and the art of blowing the old traditional form was gradually lost [8]. As one of the earliest references to distilled alcohol is found in the writings of Magister Salernus[4] [40], it is believed by some that he may have pioneered the cooling of the *solen* to effect condensation outside the still head [39].

As glass components became more common, more versatile approaches to cooling were investigated. These ideas culminated in the "wormcooler" or cooling

---

[4] Salernus (d. 1167) was a physician of the School of Salerno and lived at Salerno between 1130 and 1160. His writings included a summary of pathology and therapeutics [3, 39–41].

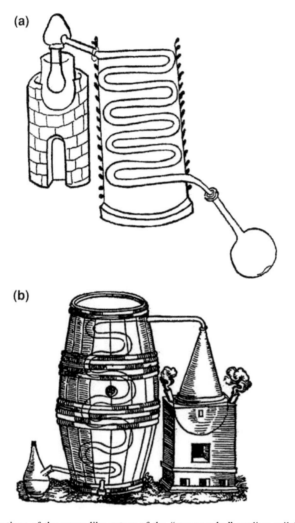

Fig. 5.8 Illustrations of the worm-like nature of the "wormcooler" cooling coil from, Philipp Ulstadt's *Coelum Philosophorum seu De Secretis Naturae Liber*, 1525 (**a**) and Walter Ryff's *Neu Gross Destillierbuch*, 1556 (**b**)

coil, which led the cooling tube through a tub of water for more efficient cooling of the delivery tube [41]. The earliest illustration of this method was given in 1420 by John Wenod, a physician in Prague, as shown in Fig. 5.7 [8, 42].

From the name "canale serpentinum" given to this cooling coil, it was thought that this tube wound worm-like through the cooling trough as shown by many later pictures (Fig. 5.8) [8, 43]. This idea was introduced in the late thirteenth century in

Fig. 5.9 The retort (Ref. [6],
Reproduced with permission
of the Division of the History
of Chemistry of the American
Chemical Society)

the writings of Taddeo Alderotti[5] of Florence (Thaddeus Florentinus, 1223–1303)
who is thought to be its inventor [3, 8]. Through the use of his "canale serpentinum"
run through a cooling trough with a regular supply of fresh cooling water, it is
thought that it was possible for Thaddeus to obtain easily 90% alcohol by
fractional distillation [3, 8, 41].

The impact of the glass industry on still evolution was evident not only in the
development of new, improved glass-based components such as the wormcooler,
but by the eventual move away from the use of even simple earthenware components to new stills fabricated completely from glass. As all-glass stills become
more common, these were eventually blown or cast in one piece. This new type of
distilling apparatus was called the *retort* (from Latin *retortus*, "bent back") and
was introduced in the early fourteenth century (Fig. 5.9). Distillation in a retort is
often referred to as *destillatio ad latus* or a "side-wards distillation" [8].

Two later designs to increase effective cooling, the *Moor's head* and the
Rosenhut, focused on the still-head instead of the solen. The Rosenhut (literally
"rose-hat") is thought to be the earlier of the two as it was illustrated in its fully
developed form in 1478 [8]. It is a high conical alembic cooled by air and is very
common in early apparatus used for making liqueurs [8, 30]. It was typically fitted
to a wide-mouthed cucurbit and must have been built with an inner rim to collect
the distillate, although this has never been shown in illustrations [8]. While the
majority of distillation apparatus were made of glass by this point, the Rosenhut
was frequently made from metals such as lead and copper as the high thermal
conductivity of the metals resulted in superior air cooling [8, 30].

The *Mohrenkopf* or "Moor's head" enclosed the still-head in a basin or container which was filled with cooling water, as shown in Fig. 5.10. The Moor's head
was typically glass (although pottery was sometimes used) and is thought to be an
invention of the later fifteenth century [8, 30]. From its name, it is believed that its
development involved influence from the world of Islam. However, it has also

---

[5] Better known as Thaddeus Florentinus, Alderotti was born in Florence in 1223. He started
teaching medicine at the University of Bologna in 1260 and thanks to his efforts the city
authorities granted medical students and teachers the legal status enjoyed by students and teachers
of law. A follower of Hippocrates, he was an author on anatomy and medicine and wrote on the
medicinal value of alcohol in his *De virtutibus aquae vitae*. He died in Bologna, although the year
of his dead is disputed, with reports of both 1303 and 1295 [3, 44].

Fig. 5.10 Illustration of the
Moor's head from
Hieronymus Brunschwyck's
*Liber de arte distillandi de
Compositis*, 1512

been suggested that the use of the term "Moor" could refer to someone from the "further Indies" and may have been influenced by the Chinese still, which also utilized a water-cooled head [41].

By this point, glass was by far the preferred medium for chemical glassware, particularly that of distillation equipment. More importantly, sixteenth-century authors such as Hieronymus Brunschwig (1450–1512)[6] and Conrad Gesner (1516–1565)[7] specified not only glass distillation components, but preferably those of Venice [45, 46]. Brunschwig even stated that the distillation vessels [45]:

..must be made of venys [Venetian] glasse bycause they shoulde the better withstande the hete of the fyre.

From such writings it was clear that specialists in Venice and Murano designed and made specific glass apparatus for the practitioner of the chemical arts.

---

[6] Brunschwig (also given as Brunschwyck, Braunschweig, or Brunschwijg) was a German surgeon and a native of Strassburg. Most believe him to have been born ∼ 1450, although others give earlier dates of 1430 or 1440. He was descended from the local Sauler (or Saler) family and studied medicine at Bologna, Padua and Paris. He is most well-known for his works on the art of distillation, most importantly *Liber de arte distillandi de simplicibus* (published May 8, 1500, commonly known as his *Small book of distillation*) and *Liber de arte distillandi de Compositis* (published February 23, 1512, commonly known as his *Large book of distillation*). He died at the age of 60, either the end of 1512 or the beginning of 1513. His death has also been reported as 1533 or 1534 [47–49].

[7] Gesner, often referred to as Evonymus Philiater, was born in Zurich on March 26, 1516. Due to his father's death, he found himself destitute at an early age and had to take an assistant's position in Strassburg. In 1535, he took a teacher's post at Zurich and studied medicine, before becoming a professor of Greek philology at Geneva in 1537. He finally finished his medical study at Basel in 1541 and then returned to Zurich as a private physician and professor of physics and natural history. He was a voluminous writer and died on December 13, 1565 [46, 48].

References

1. Neri A, Merrett C (2003) The world's most famous book on glassmaking, the art of glass. The Society of Glass Technology, Sheffield, pp 45–46
2. Taylor FS (1945) The evolution of the still. Ann Sci 5:185–202
3. Holmyard EJ (1990) Alchemy. Dover Publications, New York, pp 47–54
4. Talyor FS (1992) The Alchemists. Barnes & Noble, New York, pp 39–46
5. Macfarlane A, Martin G (2002) Glass: a world history. University of Chicago Press, Chicago, p 16
6. Rasmussen SC (2008) Advances in 13th century glass manufacturing and their effect on chemical progress. Bull Hist Chem 33:28–34
7. Philips CJ (1941) Glass: the miracle maker. Pitman Publishing Corporation, New York, pp 43–44
8. Forbes RJ (1970) A short history of the art of distillation. E. J. Brill, Leiden, pp 76–86
9. Dimbleby V, Turner WES (1926) The relationship between chemical composition and the resistance of glasses to the action of chemical reagents. Part I. J Soc Glass Technol 10:304–358
10. Turner WES (1926) The physical properties of glasses. The relationship to chemical composition and mode of preparation. J Chem Soc 129:2091–2116
11. Philips CJ (1941) Glass: the miracle maker. Pitman Publishing Corporation, New York, pp 55–57
12. Kunicki-Goldfinger JJ (2008) Unstable historic glass: symptoms, causes, mechanisms and conservation. Rev Conserv 9:47–60
13. Who was W.E.S. Turner? http://turnermuseum.group.shef.ac.uk/biography.html. Accessed 6 Dec 2011
14. Charleston RJ (1980) Our forefathers in glass. Glass Tech 21:27–36
15. Dimbleby V, Muirhead CMM, Turner WES (1922) The effect of magnesia on the resistance of glass to corroding agents and a comparison of the durability of lime and magnesia glasses. J Soc Glass Technol 6:101–107
16. English S, Turner WES (1927) Relationship between chemical composition and the thermal expansion of glasses. J Am Ceram Soc 10:551–560
17. Philips CJ (1941) Glass: the miracle maker. Pitman Publishing Corporation, New York, pp 89–90
18. Kurkjian CR, Prindle WR (1998) Perspectives on the history of glass composition. J Am Ceram Soc 81:795–813
19. Verità M (1985) L'invenzione del cristallo muranese: una verifica analitica delle fonti storiche. Rivista della Staz Sper del Vetro 15:17–29
20. Verità M, Zecchin S (2008) II - Indagini analitche di materiali vitrei del IV-XV secolo del territorio Veneziano. Rivista della Staz Sper del Vetro 38:20–32
21. Cummings K (2002) A history of glassforming. A & C Black, London, pp 102–133
22. Sarton G (1947) Introduction to the history of science. The William & Wilkins Co., Baltimore, vol III, part I, pp 170–173
23. Jacoby D (1993) Raw materials for the glass industries of Venice and the Terraferma, about 1370–about 1460. J Glass Studies 35:65–90
24. Turner WES (1956) Studies in ancient glasses and glassmaking processes. Part V. Raw materials and melting processes. J Soc Glass Technol 40:277T–300T
25. Moretti C (2001) Le materie prime dei vetrai veneziani rilevate nei ricettari dal XIV alla prima metà del XX secolo. II Parte: elenco materie prime, materie sussidiarie e semilavorati. Rivista della Staz Sper del Vetro 31:17–32
26. Moretti C (1985) Glass of the past. CHEMTECH 340
27. Mass JL, Hunt JA (2002) The early history of glassmaking in the Venetian lagoon: a microchemical investigation. Mater Res Soc Symp Proc 712:303–313

28. Macfarlane A, Martin G (2002) Glass: a world history. University of Chicago Press, Chicago, p 183
29. Forbes RJ (1970) A short history of the art of distillation. E. J. Brill, Leiden, pp 21–23
30. Ihde AJ (1964) The development of modern chemistry. Harper & Row, New York, pp 13–17
31. Bolas F (1929) The Florence flask. Pharm J 123:196–197
32. Philips CJ (1941) Glass: the miracle maker. Pitman Publishing Corporation, New York, p 384
33. Sarton G (1931) Introduction to the history of science. The William & Wilkins Co., Baltimore, vol II, part II, p 767
34. McCray WP, Kingery WD (1988) Introduction: toward a broader view of glass history and technology. Ceram Civiliz 8:1–13
35. Holmyard EJ (1956) Alchemical equipment. In: Singer C (ed) A history of technology, vol 2. Clarendon Press, Oxford
36. Stillman JM (1924) The story of early chemistry. D. Appleton and Co., New York, p 151
37. Roger F, Beard A (1948) 5,000 years of glass. J. B. Lippincott Co., New York, p 233
38. Forbes RJ (1970) A short history of the art of distillation. E. J. Brill, Leiden, p 114
39. Gies F, Gies J (1994) Cathedral, forge, and waterwheel. Technology and invention in the Middle Ages. HarperCollins Publishers, New York, p 163
40. Forbes RJ (1970) A short history of the art of distillation. E. J. Brill, Leiden, pp 57–58
41. Gwei-Djen L, Needham J, Needham D (1972) The coming of ardent water. Ambix 19:69–112
42. Partington, JR (1961–1970) A history of chemistry, vol 2. Martino Publishing, Mansfield Center, p 266
43. Liebmann AJ (1956) History of distillation. J Chem Educ 33:166–173
44. Forbes RJ (1970) A short history of the art of distillation. E. J. Brill, Leiden, pp 60–61
45. Anderson RGW (2000) The archaeology of chemistry. In: Holmes FL, Levere TH (eds) Instruments and experimentation in the history of chemistry. MIT Press, Cambridge
46. Forbes RJ (1970) A short history of the art of distillation. E. J. Brill, Leiden, pp 112, 120–124, 175–176
47. Forbes RJ (1970) A short history of the art of distillation. E. J. Brill, Leiden, pp 109–110
48. Partington JR (1961–1970) A history of chemistry, vol 2. Martino Publishing, Mansfield Center, pp 80–84
49. Tubbs RS, Bosmia AN, Mortazavi MM, Loukas M, Shoja M, Gadol AAC (2011) Hieronymus Brunschwig (c. 1450–1513): his life and contributions to surgery. Childs Nerv Syst. doi: 10.1007/s00381-011-1417-x

Chapter 6
Impact on Society and Its Effect on Chemical Progress

As presented in the previous chapter, the Venetian advancements in glassmaking led to its application in laboratory apparatus and development of laboratory glassware. While in many cases, the new laboratory glassware followed older designs previously fabricated from inferior materials, the quality and clarity of the Venetian glass also led to completely new and important objects such as lenses. In fact, some have stated that the most important long term consequence of clear glass manufacture was its development as a thinking tool through the production of mirrors, lenses, and eyeglasses [1]. Through its application in eyeglasses, glass has corrected and helped preserve our eyesight, and its use in the telescope, microscope, spectrometer, and other optical instruments has widened and deepened our ability to see that which is very small or far away [2]. This final chapter will outline the application of glass to the development of such additional critical glass objects and instruments, along with a discussion of how the developments in both the previous and current chapter impacted both society and progress in the chemical sciences.

6.1 Glass Mirrors

Various ancient texts mention mirrors silvered with lead, but such glass mirrors were always limited to very small sizes due to a number of factors. One such critical factor was the lack of techniques for producing flat, smooth glass that was still clear and relatively thin. Another problem was the fact that the application of a hot layer of lead onto glass typically resulted in thermal shock and cracking or breaking of the underlying glass substrate [3].

Roman glassmakers knew how to make glass mirrors, yet metal mirrors were thought to be preferred [4]. Early Roman glass mirrors were produced by blowing a hollow ball of glass into which molten metal was poured and then removed. The

S. C. Rasmussen, *How Glass Changed the World*,
SpringerBriefs in History of Chemistry, DOI: 10.1007/978-3-642-28183-9_6,
© The Author(s) 2012

metal used for this process has been said to be lead or tin, with silver in more rare cases [3, 4]. Mirrors produced in this fashion were never larger than what could be cut from the glass ball and thus resulted in only small, curved mirrors. In addition, the glass needed to be fairly thin and regular in order to resist the heat of the applied metal [3].

The production of larger flat mirrors was the result of a newer technique for the production of window glass. This technique, referred to as "cylinder glass", produced panes of even thickness which were glossy on both sides [3, 5, 6]. This method was first documented by Theophilus in the twelfth century and remained in use through the nineteenth century. This technique consisted of blowing a cylinder of glass, which was then opened at both ends. The resulting tube was then split longitudinally, reheated and opened out flat [5, 6]. By backing such clear glass panels with metallic leaf, flat mirrors could be made. The innovation of metallic leaf, rather than molten metal, is credited to the Venetians [7].

In the thirteenth century, the Venetians started to use a slow grinding process originally developed for the polishing of jewels in an attempt to manufacture highly polished mirrors [8]. The problem was that in order to allow the polishing and grinding needed to create a distortion-free surface, the mirror glass had to be made thicker than possible using conventional methods for the fabrication of windows. As a solution, panels of the desired thickness were typically produced by ladling as much glass as possible onto a flat, metal surface, while simultaneously attempting to smooth and flatten it. The glass sheet obtained then had to be painstakingly ground and polished before a reflecting metal foil was fixed to one surface. This process produced very high quality mirrors, but made them pro-hibitively expensive [9].

Sometime during the fourteenth and fifteenth centuries, a superior method of coating glass with a tin-mercury amalgam was developed. The exact date and location of the discovery is unknown, but credit is again typically given to the Venetian glassmakers. By the sixteenth century, Venice had become a center of mirror production using this new technique and was thought to produce the purest mirrors in the world [3].

The overall development of good glass mirrors covered the whole of Western Europe, although they were only common and of high quality starting in the thirteenth century. Mirrors were a crucial feature in the development of the sciences of optics and their applications had significant effects on the developing sciences of chemistry and astronomy. It has been said that without mirrors, the Renaissance and the Scientific Revolution might not have occurred [10].

6.2 Eyeglasses

Once the polishing technique the Venetians utilized in mirror glass became more common, the manufacturing of spherical glass surfaces became much easier. The first application of the resulting polished lenses was eyeglasses [8]. Credit for their invention has been attributed to various people over time, including the Dominican Friar Alessandro Spina of Pisa [11], Salvino del' Armani of Florence, and Roger Bacon [12], who had proposed in his *Opus Majus* the use of a lens (a segment of a glass sphere) to help the elderly afflicted with weak eyesight [13]. However, available historical evidence has shown all of these to be false attributions and the specific inventor of eyeglasses is still unknown [11, 12]. Available sources point towards their appearance in Italy shortly after 1286 [1, 8, 11, 12, 14], but prior to 1292 [14] and they were most likely developed in Pisa (although Venice has also been supported by some as the site of discovery) [12–14]. Once introduced, their use spread rapidly throughout Europe [1].

The earliest known depiction of eyeglasses is a portrait of Dominican cardinal Hugh of St. Cher (ca. 1200–1263) painted by Tommaso de Modena in 1352 [2, 12–14, 16]. In a bit of artistic freedom, de Modena included eyeglasses in the portrait, even though Hugh died more than 20 years before their invention. This practice was commonplace as glasses were often associated with wisdom or learning [13, 14] and points to the fact that eyeglasses were familiar by the mid-fourteenth century [12].

The production of eyeglasses was facilitated by the glassmakers' mastery of the making of uniform, clear glass necessary for a good supply of quality lenses [12, 15], and it has been speculated that the original inventor was most likely an experienced glassworker [11, 16]. By 1300, eyeglasses were being produced by Venetian glassmakers and were repeatedly referenced in guild regulations during the first two decades of the fourteenth century [12, 13, 16]. In fact, Venice became such an important center for the production of eyeglasses that Venetian spectacle makers left the existing glassmakers' guild to form their own guild in 1320 [13]. While glassworks and the production of eyeglasses became established in other regions, the glass of Murano was considered to be of superior quality and a more suitable substance for the grinding of quality lenses. As a result, Murano glass continued to be imported into these regions even after independent glassworks had been established [13]. By the fifteenth century, Florence was considered a leading center for the production of high-quality eyeglasses [16]. However, while these glasses were ground and assembled in Florence, it is unclear if the lenses were ground from locally produced glass or glass imported from Murano.

Modern eyeglasses with temples (the side-pieces that extend over and/or behind the ears to hold them in place) did not become commonplace unit the nineteenth century [13]. The earliest spectacles were comprised of two separate lenses and frames, held together with a simple rivet (Fig. 6.1a) [13, 14]. The frames were constructed from a variety of materials including metals, bone, and wood. Serrations on the inside of the frames beneath the rivet could then be used to prevent the spectacles from falling off the nose. Such spectacles could not have been comfortable for long periods of use, however, and some paintings have shown the reader holding the riveted junction between the thumb and fore-finger with the hand and wrist held just clear of the brow [13]. These initial spectacles utilized convex lenses (Fig. 6.1b) [1, 15, 16], thus improving vision for the farsighted and used primarily for reading [12, 13]. Concave lenses, for the nearsighted, were more difficult to work and did not arrive until the mid-fifteenth century [1, 14, 16]. Even after the introduction of

Fig. 6.1 Early eyeglasses design (**a**) and convex versus concave lenses (**b**)

concave lenses, however, the majority of glasses were still commonly used for reading [14].

Eyeglasses have been stated to be one of mankind's most beneficial material inventions. Without eyeglasses, people born with poor vision would be illiterate or have insufficient vision for a skilled trade. Even most people born with normal vision typically lose the ability to focus by their mid-40s [1, 13]. As a consequence, the single invention of eyeglasses nearly doubled the intellectual life span of the average person from the thirteenth century on, resulting in a significant impact on society as a whole.

6.3 Microscope and Telescope

It seems likely that the increasing use of eyeglasses, resulting in a general familiarity with lenses, provided a rich environment for the development of both the microscope and the telescope [2]. Both instruments seem to have originated in the Netherlands around the year 1600, with knowledge of their discoveries diffusing throughout Western Europe [17, 18]. As the glassmaking knowledge of Venice started to spread across Europe in the sixteenth century, the Netherlands was one of the important centers to which these skills and knowledge were transmitted. It has been reported that the fabrication of cristallo reached Antwerp in 1537 and a Venetian-founded mirror factory was established there in 1541 [19].

It is believed that the microscope was the first of these two instruments produced. However, as with the case of eyeglasses above, the origin of the compound microscope is still a matter of debate. Credit for this invention is typically given to Zacharias Jansen[1] (1580–1638), with this accomplishment generally dated in the 1590s [2, 8, 20]. Due to Zacharias' young age, it is believed that his father,

---

[1] Alternate forms include Sacharias Jansen, Zacharius Jansen, Zacharias Janssen, and Zacharias Janssens [2, 8, 20].

Johannes or Hans,[2] must have played a significant role in the microscope's development. Both father and son worked together as spectacle makers in Middleburg, Holland, and thus had a working knowledge of lenses and their fabrication [8, 20].

The compound microscope contains at least two lenses, thus distinguishing it from the simple microscope, which contains a single lens like that of the magnifying glass. The microscope credited to Jansen had two lenses combined in a segmented tube in which the magnification was varied by changing the length of the tube and thus the distance between the lenses. In principle, the combination of two lenses should produce a clearer image of greater magnification. In practice, however, the simple microscope was able to compete with the compound microscope for much of the seventeenth century. It has been suggested that this was partially due to the low quality of many available lenses. However, a greater limitation was the fact that the combination of lenses distorted the image and produced colored halos, thus reducing the advantage over a single lens [18].

The Jansens are not known to have published any observations through their microscope and the names most associated with early microscopy are Robert Hooke (1635–1703) and Antony van Leeuwenhoek (1632–1723) [2, 8, 21]. In his 1665 book *Micrographia*, Hooke described the manufacture of the first high-power microscope lenses [8, 17, 21]. By drawing the glass into long threads and then melting the tips into very small balls, high quality lenses of very short focal length could be produced [2, 8]. While much of Hooke's work applied these new lenses to simple microscopes, he also describes a compound microscope consisting of a tube of 6–7 inches with an object lens and two eye lenses [8]. Hooke's book then inspired van Leeuwenhoek who began to experiment by making microscopes of the kind Hooke described [21]. Leeuwenhoek utilized high quality simple microscopes and gave the first accurate description of the red blood corpuscles, investigated the structure of the teeth, the lens of the eye, and other physiological objects. He was also the first microscopist who gave a description of bacteria [8]. It has been stated that the glass utilized for these early microscopes were all made from cristallo [9]. While there does not seem to be any direct evidence of this, Hooke does specify Venetian glass in his construction of the high-power lenses [17].

Unfortunately the origin of the telescope is really no clearer than that of the microscope, with three rival claimants for its discovery [18]. The Dutch spectacle-maker Johann (or Hans) Lippershey[3] (1570–1619), along with the Jansens discussed above, both applied for patents in 1608. Of these, Lippershey is usually considered to have the best claim and is most often given credit for the discovery of the telescope [2, 18, 20]. While there have been references to earlier instruments giving optical magnification, it is not believed that the telescope was constructed before 1608 [18]. Lippershey set lenses of his own grinding into a tube, allowing the viewing of distant objects. Hearing of its construction, Galileo Galilei (1564–1642) then built his own

---

[2] Hans is the diminutive form of Johannes, but can also be an independent given name.
[3] Surname sometimes given as Lippersheim [20].

telescope in 1609 [18, 20]. Although it was a crude instrument compared to modern analogues, Galileo was the first to turn his telescope on the heavens and published his observations in 1610. In the process, he studied the sun, moon, and the planets and was able see the four giant moons of Jupiter for the first time [2, 18]. More importantly he found that the heavenly bodies were neither perfect nor unchanging, thus leading to an overturning of Aristotle's celestial physics.

A common question of scholars is why it took so long for the microscope or telescope to be discovered following the advent of lenses for eyeglasses? For the microscope, this is unclear. For the telescope, however, the delay was at least partially due to the fact that its construction required the combination of a convex with a concave lens at a proper distance to give an erect image [16]. As discussed above, concave lenses were more difficult to grind and did not arrive until the mid-fifteenth century [1, 14, 16]. As such, the availability of high quality concave lenses alone would have caused a necessary delay in the telescope's fabrication. However, it is also believed by some that it took this period of time for lenses to move from a simple practical device for the improvement of eyesight to something of academic interest and general study [16].

Without the microscope, the modern discoveries of the origin of diseases, and their control and cure, could not have been affected. Likewise, our knowledge of the structure of a wide variety of materials, from metals, rocks, and minerals to cotton, wool, and silk fibers, have all been due to the microscope [2, 21]. At the same time, telescopes revealed the wonders of the heavens, as well as allowing us to span great distances on land and at sea [2].

6.4 Thermometer and Barometer

The properties and versatility of glass allow the development of additional critical instruments such as the thermometer and barometer, both developed in Italy in the early seventeenth century [10, 22]. Again priority for the discovery of the thermometer is not clear and four names have been referenced in relation to its invention: Santorio Santorii[4] (1561–1636), Galileo Galilei, Robert Fludd (1574–1651), and Cornelius Drebbel (1572–1633) [23]. Of these, the best cases can be made for either Santorio or Galileo.

The earliest known account of an instrument for measuring heat and cold was published by Santorio in Part III of his *Commentaria in artem Medicinalem Galenis* dated 1612 [23]. In this relevant passage, he refers to it as a "very ancient instrument". However, in a letter dated May 9, 1613, he refers to Galileo as its inventor. To further confuse things, the first illustration of the thermometer was published in 1617 by Giuseppe Biancani, who credits Santorio with its invention [23].

---

[4] Also known as Santorio Santorio or Sanctorious [M,N].

Galileo's claim as inventor of the instrument depends entirely on the assertions of his friends as none of his letters or works known to be written before 1612 make any allusion to the thermometer or any similar instrument [23]. However, the very fact that Santorio gives credit to Galileo gives his claim significant credence. As such, it is generally held that Galileo seems to have been the initial inventor at some period between 1592 and 1603 [18, 23, 24], while Santorio gives the first written record of the invention, published or unpublished. Even so, Galileo certainly gave little or no attention to the use and development of the thermometer, while Santorio seems to have considerably improved it, adopting both a fixed scale and making meteorological observations [23].

The air thermometer of Galileo and Santorio consisted of a long, thin glass tube terminated with a glass bulb. This was inverted over a container of water such that some of the water was sucked up the tube. When the air in the bulb was heated, it expanded and drove the water down the tube. Thus, the height of the water in the tube could be related to the temperature of air in the bulb [18]. However, the initial air thermometer had serious disadvantages. Not only was it clumsy and difficult to move around, but it was also greatly affected by barometric pressure and thus gave results of limited accuracy [18, 24].

The thermometer was then made a bit more practical a few years later by Ferdinand II, Grand Duke of Tuscany (1610–1670). He modified the thermometer by filling the bulb with liquid and sealing the end of the tube. In this case, the liquid was now the thermometric material and the thermometer was essentially independent of barometric pressure [24]. It was not until the early eighteenth century that even more accurate thermometers based on the expansion of mercury were being produced by such instrument makers as the German G. D. Fahrenheit (1686–1736) [18]. Not only did the thermometer allow a quantitative determination of temperature, but its applications to medicine helped invest the discipline with much increased power [10, 22].

The dependence of the thermometer on pressure led to the invention of an instrument similar to the air thermometer, the barometer. This was invented by Evangelista Torricelli (1608–1647), a pupil of Galileo, in 1643 [18]. Here, an evacuated glass tube is partially filled with mercury and inverted over a mercury reservoir. This results in a column of mercury in the tube, with the remaining headspace a vacuum, and the height of the mercury would depend on the pressure exerted by the atmosphere on the mercury in the reservoir.

6.5 Alcohol

In addition to the direct applications discussed so far, the availability of glass-based apparatus resulted in other critical discoveries that forever changed the face of science and society. Perhaps one of the greatest of such cases was utilization of the improved glass distillation equipment for the isolation of alcohol from wine. The successfully isolated distillate then found a variety of uses—as a solvent, preservative, and the basis of brandy, at first taken medicinally, later recreationally [15].

Although alcoholic drinks from fermentation have been known since ~ 8000 BCE, alcohol as an independent species was not recognized until the beginning of the twelfth century. Available evidence clearly shows that alcohol was discovered in ~ 1100 AD, most likely at the School of Salerno, an important medical school [25–30]. This assignment is strengthened by the fact that one of the earliest direct recipes for isolation of alcohol is contained in the writings of Magister Salernus. Earlier texts from Salerno discuss the preparation of "beneficial waters" by distillation, but his writings were the first to directly mention alcohol [27]. The reason for the late discovery of alcohol was partly due to the long preheating period coupled with inefficient cooling during distillation. However, another factor was that even the most refined alcoholic distillate separated by the early stills contained so much water that it would not burn, thus making it difficult to differentiate from normal water [25, 27].

The secret of the success after 1100 was not just better cooling and fractional distillation, but also the addition of various salt substances (NaCl, potassium tartrate, K_2CO_3, etc.) which absorbed part of the water, thus increasing the concentration of alcohol and making it easier to separate by distillation [27]. These early solutions distilled from wine-salt mixtures were referred to as *aqua ardens* (burning water) or *aqua flamens* (flaming water) and had such low alcohol content that they burned without producing noticeable heat [25, 26, 30–32]. The combination of more efficient cooling with the earlier salt additions produced alcoholic distillates containing less than 35% water and absolute alcohol could be obtained after several fractional distillations [27].

Monks had long produced wines for sacramental purposes, and were some of the first distillers in the west. Initially, they thought alcohol was the *quintessence*, the fifth element that made up the heavens, and they studied its properties intensively. Alcohol looked like water, but burned with a blue, gemlike flame, although all knew that the nature of water was to extinguish fire, not to burn. In addition, when consumed, it produced a very pleasing intoxication. An alchemical text ascribed to Raymond Lull described alcohol as follows [33]:

> The taste of it exceedeth all other tastes, and the smell of it all other smells.

The separation of alcohol from wine was analogous to the separation of the soul from an impure body and for this reason was thought to be the "spirit" of the wine and the remaining residue was called the *caput mortum* (dead body). Absolute or very strong alcohol was also named *aqua vitae* (water of life) by Arnold of Villanova [27, 30, 34] and the name still survives in the modern words *aquavit* (Scandinavian), *eau-de-vie* (French), *whiskey* (Scottish), and *vodka* (Slavic). Arnold of Villanova said the following on his choice of the name *aqua vitae*:

> This name is remarkably suitable, since it is really a water of immortality. It prolongs life, clears away ill-humours, revives the heart, and maintains youth.

Our modern term *alcohol* was not used to refer to these distillation products until the sixteenth century, when it was introduced by Paracelsus [26]. The word alcohol comes from the arabic *al-kohl*, which originally referred to finely powdered antimony trisulphide [26, 27]. However, it came to be used to refer to any very fine powder [27]. This meaning was then further extended to mean the most subtle part of something [26, 27] and was thus generally used for any substance attenuated by pulverization, distillation, or sublimation [35]. By the sixteenth century, aqueous solutions distilled from wines were referred to as *alkohol vini* (i.e. the subtle part of wine) and eventually the *vini* was dropped [26].

By 1288, the study of alcohol had perhaps gone a bit too far, because the Dominican provincial chapter at Rimini declared it forbidden for brethren to possess the instruments by which they make the water called *aqua vitae* [31]. This was at least partially because its production was associated with apothecaries, who were the first to produce alcohol on a large scale [27], and the secular nature of the apothecary trade was considered inappropriate or beneath such men of the Church [25]. Nevertheless, other monks continued to produce alcoholic beverages such as the oldest known liqueur *Benedictine*, invented by Dom Bernardo Vincelli in 1510 [31]. This liqueur is made through the extraction and distillation of a variety of herbs with alcohol. A related liqueur *Chartreuse* was later made by the Carthusian monks [32].

Alcohol became a common reagent of the laboratory where it was used as a powerful solvent. Not only could it solubilize most salts and other water-soluble substances, but it also dissolved many organic materials not soluble in water, such as fats, resins, and essential oils. In that respect, it was the first broadly applied solvent that could be used to solubilize less polar species. For example, it was the first liquid that could be used to extract the volatile aromatic substances from plants [31, 32]. This greatly expanded the number of possible useful solutions available to the practicing alchemist and in many ways changed the focus of chemical investigations.

At the same time, alcohol began to be used as a medicine in the mid-thirteenth century. The Italian physician Thaddeus Florentinus and the Franciscan Vitalis de Furno (ca. 1260–1327)[5] have been credited with the earliest application of medicinal alcohol [26]. It was generally reasoned that purified alcohol would in turn purify the patient from illness and thus by 1288, alcohol as a medicine was in general use [31]. Its effect on those that consumed it could be clearly seen and its effect on the failing powers of the aged led to its use as a medicine against old age. The fact that alcohol exhibited the property of preserving organic matter from putrefaction probably also helped support the idea that it would preserve the human body. In addition, the belief that alcohol was the quintessence gave reason

---

[5] Known also as Vital du Four, Vital du Fourca, and Joannes Vitalis. He was a Franciscan theologian and scholastic philosopher who played a prominent role in the controversy over the Franciscan conception of *usus pauper*. He was born at Bazas in Aquitaine, about 60 km southeast of Bordeaux. He entered the Franciscan order at an early age and went to study theology at Paris from 1285 to 1291. He taught at Montpellier from 1292 to 1296, after which he was transferred to the University of Toulouse. He was made cardinal-priest by Pope Clement V in 1312 and became bishop of Albano in 1321 [36].

for the presumption that it would prove to be the most perfect of medicines [31]. In a more practical sense, washing wounds with alcohol cleansed them and killed some microorganisms. In addition, administering alcohol to the patients relieved pain and made them relaxed, perhaps even happy, thus allowing the body a chance to heal itself [29, 30]. In 1347–1351, Europe was in crisis as it dealt with an epidemic of plague, the well-known Black Death, and alcohol was the primary medicine that could give relief [29, 30, 32]. As a cure for the plague, alcohol was ineffective, but at least it made the patient who drank it feel more robust. No other known treatment could even do that much [29]. Alcohol was also prescribed for cases of typhoid fever, diarrhea, and other similar diseases [32]. By the mid-fourteenth century, the medicinal and preservative properties of pure alcohol became the backbone of the writings of such authors as Arnald of Villanova[6] and John of Rupescissa,[7] and it was soon widely recommended as a universal remedy [25, 31, 32].

6.6 Mineral Acids

Another class of important chemical species that changed chemical practices was the mineral acids. While earlier practitioners were well acquainted with the vitriols (i.e. metal sulfates) [37] and their calcination products, it is believed that the acid vapors had not been condensed prior to the thirteenth century. It has been suggested that the new still design of the retort may have been important in the

---

[6] Arnaldus de Vill Nova, or Villanovanus. Born near Valencia about 1240 (c. 1234–1250), he studied medicine at Naples, traveled extensively, living in Paris, Montpellier, Barcelona, and Rome. He was a Catalan physician, alchemist, astrologer, and diplomat, and was a professor at Montpellier at least up to 1309. He translated medical works from Arabic into Latin and had some knowledge of Greek and Hebrew. He realized the value of natural science and suggested that it should be given more importance in education. He was a famous medical practitioner, who was consulted by kings and popes, and is considered by some to be one of the most extraordinary personalities of medieval times. He also had difficulties with the French Inquisition, first in 1299 and again in 1304. He had a large number of writings ascribed to him, most of them dealing with medical subjects and he was one of the first Latin writers to insist upon the virtues of alcohol. Other works dealt with chemistry, astrology, magic and theology. Most of these writings are very short and many are apocryphal. He died at sea towards the end of 1311 while on the way from Naples to Genoa, where he was buried [30, 32, 41].

[7] John de Rupescissa or John of Roquetallaide (d. 1362). An often cited author about which little is really known, he lived in the middle of the fourteenth century and was a tertiary member of the Franciscan order. He was known to his contemporaries for his apocalyptic preaching, for which was often imprisoned. He was a Catalan, but a significant number of his books were written in Latin. He studied in Toulouse for five years before entering the Franciscan monastery at Orléans, where he continued his studies for five more years. He was imprisoned for the first time in 1345, and again in 1346, 1349 and 1356. His principle work, *De consideratione quintae essentiae (On the consideration of the fifth essence),* consists of two parts—canons and remedies; the backbone of which seems to be the medicinal and preservative properties of pure alcohol. The idea of associating alcohol with the quintessence is frequently credited to Rupescissa [32, 42].

preparation of the mineral acids, as its one-piece design would have been beneficial for such corrosive compounds [38]. Before the introduction of the retort, the various pieces of the still were fixed together using a *lute*. The lute was a material of adequate plasticity that was used to seal the various joints, particularly between the curcurbit and the still-head. A variety of materials have been reported to be used for this purpose, including fat, wax, lime mixed with egg white, and clay mixed with oil [39, 40]. Needless to say, attempts to distill acidic vapors would have resulted in reactions between the vapors and the lute, which could have resulted in additional problems. Even ignoring potential complications with the lute, glass or other still materials used for such isolations required good chemical resistance, which may have been a factor limiting an earlier discovery.

Although references to nitric and sulfuric acid have been found in Byzantine manuscripts of the late thirteenth century [38], the earliest known recipe for the preparation of nitric acid comes from the sixteenth century text *De inventione veritatis* of the Pseudo-Geberian corpus [37]. Nitric acid was generally prepared by the distillation of mixtures of nitre ($NaNO_3$) or saltpeter (KNO_3) with either alum ($NH_4Al(SO_4)_2 \cdot 12H_2O$) or sal ammoniac ($NH_4Cl$) [37, 38, 43]. Some descriptions of these processes have denoted dry distillations, while others utilize aqueous solutions. Some recipes also call for the use of vitriol (most likely $CuSO_4$ or $FeSO_4$) [37]. In the case of dry distillations, the resulting acidic vapors would condense in the still head along with adventitious water (typically waters of hydration from the various reagents), thus producing aqueous nitric acid solutions. This acid was soon produced in large quantities as a sideline of the saltpeter industry, and by the fifteenth century Venice had become a center for its large-scale manufacture [38].

In the same way, the early history of sulfuric acid is also difficult to trace and no reliable recipe for its preparation is known prior to the sixteenth century [37]. Sulfuric acid (oil of vitriol) was commonly prepared by first "roasting" or calcining vitriol (usually green vitriol or hydrated $FeSO_4$) in an earthen vessel to produce a crude mixture of metal oxide and sulfuric acid as shown in Scheme 6.1 [37, 38, 43]. The mixture was then distilled in a glass retort to isolate the desired acid solution.

Another common method for the production of sulfuric acid was referred to as the "bell-process" [38]. This method involved the burning of sulfur with excess air and in contact with water, the chemistry of which is outlined in Scheme 6.2. The name "bell-process" referred to the tradition of burning the sulfur under a glass bell suspended over a vessel or pan of water. As the sulfur burned, the glass bell would trap the sulfur oxides in order to allow them react with the water vapor. The resulting sulfurous and sulfuric acids would thus condense on the interior of the bell to be collected in the water containing vessel beneath [38, 43].

Although the preparation of hydrochloric acid seems to have occurred at a later date and was not commonly used until the seventeenth century [38], methods for the direct preparation of nitric acid-hydrochloric acid mixtures were known in the same early period as nitric acid. The most straight-forward of these methods was the simple addition of sal ammoniac (NH_4Cl) to freshly distilled nitric acid [37]. As still referred to today, these mixtures were called *aqua regia* (royal water) as they could readily dissolve gold and other noble metals.

$$2 \; FeSO_4 \cdot nH_2O \; (s) + {}^1\!/_2 \, O_2 \, (g) \longrightarrow Fe_2O_3 \, (s) + (n\text{-}2) \, H_2O \, (l) + 2 \, H_2SO_4 \, (aq)$$

Scheme 6.1 Roasting of green vitriol to produce sulfuric acid

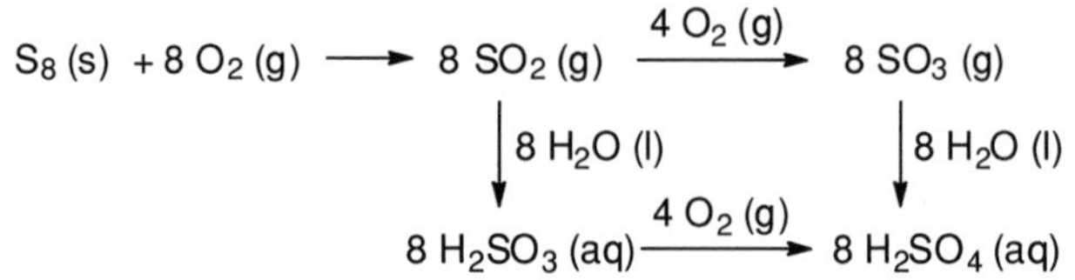

$$S_8 \, (s) + 8 \, O_2 \, (g) \longrightarrow 8 \, SO_2 \, (g) \xrightarrow{\; 4 \, O_2 \, (g) \;} 8 \, SO_3 \, (g)$$

with vertical arrows: $8 \, H_2O \, (l)$ and $8 \, H_2O \, (l)$

$$8 \, H_2SO_3 \, (aq) \xrightarrow{\; 4 \, O_2 \, (g) \;} 8 \, H_2SO_4 \, (aq)$$

Scheme 6.2 Production of sulfuric acid from the burning of yellow sulfur

The application of these acid reagents quickly changed the laboratory setting, as access to acidic solutions allowed practitioners to dissolve metals and most ores either at room temperature or in a water bath. This removed the need for enormous furnaces in special workshops, since glass vessels at workbenches were now sufficient for many processes, and entirely new classes of room temperature reactions were now possible. In addition, the reduction in laboratory equipment now needed to do meaningful study led to an enormous increase in the number of people who could do laboratory work, thus greatly accelerating the rate of progress in chemical technology.

6.7 Conclusion

Up until the discovery of organic plastics in the twentieth century [44], there was no other material that had such an impact as glass did on the world. From its early applications to beads and jewelry to its first height in the Roman Empire, it became a part of everyday household objects and opened our homes to the outside world with glass windows. However, it wasn't until the growth and eventual dominance of the glass industry in Venice and Murano during the thirteenth to eighteenth centuries that glass reached its true potential. The combination of carefully chosen reagents and new methods for their purification led to new compositions of glass with enhanced clarity and workability, as well as significantly improved chemical and thermal durability. The quality of the Venetian glasses dominated the European glass-making industry until the eighteenth century and had a direct impact on the advancement of the sciences. The ability to produce laboratory apparatus and vessels from glass allowed much greater freedom and versatility in the design of chemical glassware. Nowhere was this more evident than in the rapid and vast improvement in distillation apparatus. As a result of better stills, important materials were isolated in pure forms for the first time, most importantly alcohol and the mineral acids. The availability of these materials then greatly changed the evolving fields of both chemistry and medicine and marked the beginning of a new stage in the history of

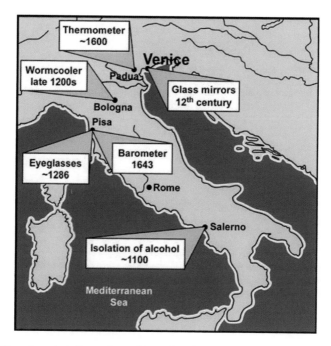

Fig. 6.2 Glass-dependent discoveries and inventions in close proximity to Venice

chemistry [27, 45]. It is at this point in history where the process of acid dissolution is combined with the process of distillation in sequential rarefication in order to produce new alchemical medications [42], which in turn eventually led to the first real beginnings of true chemistry. In fact, by the late sixteenth century the term *alchemy* became restricted to the operations of the gold-seeker, and the terms *chymia* and then *chymistry* began to be used to refer to the study of matter.

Of course, the impact of the Venetian glass was not on medicine and chemistry alone. Both glass mirrors and eyeglasses changed how everyone viewed themselves and the world around them. The commonality of eyeglasses then led to our familiarity with lenses, which in turn led to the invention of the telescope and microscope. The discovery of the telescope changed our view of the heavens and the Universe, forever changing astronomy. In a similar manner, the microscope revealed all that was too small to see and the instrument became the recognizable symbol of the biological sciences. Finally the freedom to form glass into any shape the mind can envision led to such inventions as the thermometer and barometer. The former of these is so taken for granted that it seems commonplace, but the knowledge and measurement of temperature is fundamental to everything—be it the next laboratory experiment or determining what to wear for the day. Not only is all of this dependent on the application of glass, but many required the critical combination of material strength, chemical durability, thermal durability, and optical clarity, initially all properties only found collectively in the Venetian glass. In addition, many of these discoveries and inventions occurred during the exact time period of the dominance of Venetian glass and many within Italy, some even within close proximity to Venice itself (Fig. 6.2). Without a doubt, glass has had far reaching impact and permanently changed our world. In closing, I will again turn to the eloquent words of Neri [46]:

> Wherefore for these and many other reasons you may well conclude, that Glass is one of the most Noble things which man hath at this day, for his use upon the earth.

References

1. Macfarlane A, Martin G (2002) Glass, a world history. University of Chicago Press, Chicago, pp 14, 144–150
2. Philips CJ (1941) Glass: the miracle maker. Pitman Publishing Corporation, New York, pp 373–380
3. Melchior-Bonnet S (2001) The mirror (trans: Jewett KH). Routledge, New York, pp 13–21
4. Macfarlane A, Martin G (2002) Glass a world history. University of Chicago Press, Chicago, p 16
5. Taylor M, Hill D (2001) No pane, no gain. Glass News 9:6
6. Taylor M, Hill D (2002) An experiment in the manufacture of Roman window glass. ARA Bull 13:19
7. Roger F, Beard A (1948) 5,000 years of glass. J. B. Lippincott Co., New York, pp 28–38
8. Fassin G (1934) Something about the early history of the microscope. Sci Mon 38:452–459
9. Cummings K (2002) A history of glassforming. A & C Black, London, p 122
10. Macfarlane A, Martin G (2002) Glass: a world history. University of Chicago Press, Chicago, pp 182–184
11. Rosen E (1956) The invention of eyeglasses, Part I. J Hist Med 11:13–46
12. Rosen E (1956) The invention of eyeglasses, Part II. J Hist Med 11:183–218
13. Dreyfus J (1988) The invention of spectacles and the advent of printing. The Library 10:93–106
14. Letocha CE, Dreyfus J (2002) Early prints depicting eyeglasses. Arch Ophthalmol 120:1577–1580
15. Gies F, Gies J (1994) Cathedral, forge, and waterwheel. Technology and invention in the middle ages. HarperCollins Publishers, New York, p 227
16. Ilardi V (1976) Eyeglasses and concave lenses in fifteenth-century Florence and Milan: new documents. Renaissance Quart 29:341–360
17. Macfarlane A, Martin G (2002) Glass: a world history. University of Chicago Press, Chicago, pp 84–86
18. Meadows J (1992) The great scientists. The story of Science told through the lives of twelve landmark figures. Oxford University Press, NewYork, pp 37–40, 78–79
19. Macfarlane A, Martin G (2002) Glass: a world history. University of Chicago Press, Chicago, p 23
20. Roger F, Beard A (1948) 5,000 years of glass. J. B. Lippincott Co., New York, p 239
21. Ford BJ (2001) The Royal Society and the microscope. Notes Rec R Soc Lond 55:29–49
22. Philips CJ (1941) Glass: the miracle maker. Pitman Publishing Corporation, New York, p 384
23. Talyor FS (1942) The origin of the thermometer. Ann Sci 5:129–156
24. Noyes B (1936) The history of the thermometer and the sphygmomanometer. Bull Med Libr Assoc 24:155–165
25. Gwei-Djen L, Needham J, Needham D (1972) The coming of ardent water. Ambix 19:69–112
26. Stillman JM (1924) The story of early chemistry. D. Appleton and Co., New York, pp 187–192
27. Forbes RJ (1970) A short history of the art of distillation. E. J. Brill, Leiden, pp 87–90

28. von Lippmann EO (1920) Zur geschichte des alkohols. Chem.-Ztg. 44:625
29. Vallee BL (1998) Alcohol in the western world. Sci Am 279:80–85 June
30. Liebmann AJ (1956) History of distillation. J Chem Educ 33:166–173
31. Taylor FS (1992) The alchemists. Barnes & Noble, New York, pp 98–100
32. Forbes RJ (1970) A short history of the art of distillation. E. J. Brill, Leiden, pp 60–65,91–92 and 95
33. Read J (1995) From Alchemy to chemistry. Dover, New York, p 21
34. Fleming A (1975) Alcohol, the delightful poison. Delacorte Press, New York, p 12
35. Forbes, RJ (1970) A short history of the art of distillation. E. J. Brill, Leiden, pp 47,107
36. Traver AG (2003) 126. Vital du Four. In: Gracia JJE, Noone TB (eds.) A companion to philosophy in the middle ages. Blackwell Publishing Ltd, Malden
37. Karpenko V, Norris JA (2002) Vitriol in the history of chemistry. Chem Listy 96:997–1005
38. Forbes, RJ (1970) A short history of the art of distillation. E. J. Brill, Leiden, pp 86–87, 170–174
39. Taylor FS (1945) The evolution of the still. Ann Sci 5:185–202
40. Forbes RJ (1970) A short history of the art of distillation. E. J. Brill, Leiden, p 23
41. Sarton G (1931) Introduction to the history of science. The William & Wilkins Co., Baltimore, vol II, part II, pp 893–900
42. Multhauf JP (1954) John of Rupescissa and the origin of medical chemistry. Isis 45:359–367
43. Partington JR (1961–1970) A history of chemistry, vol 2. Martino Publishing, Mansfield Center, pp 23–24
44. Strom ET, Rasmussen SC (eds) (2011) 100+ years of plastics. Leo Baekeland and Beyond. ACS symposium series 1080, American Chemical Society, Washington
45. Rasmussen SC (2008) Advances in 13th century glass manufacturing and their effect on chemical progress. Bull Hist Chem 33:28–34
46. Neri A, Merrett C (2003) The world's most famous book on glassmaking, the art of glass. The Society of Glass Technology, Sheffield, p 49

About the Author

Seth C. Rasmussen is an Associate Professor of Chemistry at North Dakota State University (NDSU) in Fargo (seth.rasmussen@ndsu.edu). He received his B.S in Chemistry from Washington State University in 1990 and his Ph.D. in Inorganic Chemistry from Clemson University in 1994, under the guidance of Prof. John D. Peterson. As a postdoctoral associate at the University of Oregon, he then studied conjugated organic polymers under Prof. James E. Hutchison. In 1997, he accepted a teaching position at the University of Oregon, before moving to join the faculty at NDSU in 1999.

Active in the fields of materials chemistry and the history of chemistry, his research interests include the design and synthesis of conjugated materials, photovoltaics (solar cells), organic light emitting diodes, the application of history to chemical education, the history of materials, and chemical technology in antiquity. As both author and editor, Prof. Rasmussen has contributed to books in both materials and history and has published more than 50 research papers. He is a member of various international professional societies including the American Chemical Society, Materials Research Society, Alpha Chi Sigma, Society for the History of Alchemy and Chemistry, and the International History, Philosophy & Science Teaching Group.

Professor Rasmussen currently serves as the Program Chair for the History of Chemistry division of the American Chemical Society and as Series Editor for *Springer Briefs in Molecular Science: History of Chemistry*.

S. C. Rasmussen, *How Glass Changed the World*,
SpringerBriefs in History of Chemistry, DOI: 10.1007/978-3-642-28183-9,
© The Author(s) 2012

Made in the USA
Lexington, KY
07 July 2016